IfColog Proceedings
Volume 2

Future Directions for Logic
Proceedings of PhDs in Logic III

Volume 1
Proceedings of URC* 2010. Undergraduate Research in Computer Science
– Theory and Applications. Student Conference
Maribel Fernández and Kathleen Steinhöfel, eds.

Volume 2
Future Directions for Logic. Proceedings of PhDs in Logic III
Jonas De Vuyst and Lorenz Demey, eds.

Future Directions for Logic
Proceedings of PhDs in Logic III

Edited by

Jonas De Vuyst

and

Lorenz Demey

© Individual author and College Publications 2012. All rights reserved.

ISBN 978-1-84890-079-0

College Publications
Scientific Director: Dov Gabbay
Managing Director: Jane Spurr
Department of Computer Science
King's College London, Strand, London WC2R 2LS, UK

http://www.collegepublications.co.uk

Original cover design by Laraine Welch
Printed by Lightning Source, Milton Keynes, UK

Contents

Author Affiliations

Editors

Jonas De Vuyst

Center for Logic and Philosophy of Science
Vrije Universiteit Brussel, Belgium
jonas.de.vuyst@vub.ac.be

Lorenz Demey

Center for Logic and Analytical Philosophy
KU Leuven – University of Leuven, Belgium
lorenz.demey@hiw.kuleuven.be

Contributors

Raphaël Carroy

HEC, Research Group in Mathematical Logic
University of Lausanne, Switzerland
raphael.carroy@unil.ch

Equipe de Logique Mathématique
Université Paris Diderot Paris 7, France

Dion Coumans

Institute for Mathematics, Astrophysics and Particle Physics
Radboud Universiteit Nijmegen, The Netherlands
d.coumans@math.ru.nl

Vincent Degauquier

Institut supérieur de Philosophie
Université catholique de Louvain, Belgium
vincent.degauquier@uclouvain.be

Kevin Fournier

HEC, Research Group in Mathematical Logic
University of Lausanne, Switzerland
kevin.fournier@unil.ch

Equipe de Logique Mathématique
Université Paris Diderot Paris 7, France

Tjerk Gauderis

Center for Logic and Philosophy of Science
Universiteit Gent, Belgium
tjerk.gauderis@ugent.be

Sam van Gool

Institute for Mathematics, Astrophysics and Particle Physics
Radboud Universiteit Nijmegen, The Netherlands
s.vangool@math.ru.nl

Jens Ulrik Hansen

Department of Philosophy/Department of Computer Science
Roskilde University, Denmark
jensuh@ruc.dk

Philipp Lücke

Mathematisches Institut
Rheinische Friedrich-Wilhelms-Universität Bonn, Germany
pluecke@math.uni-bonn.de

Alexandru Marcoci

Department of Philosophy, Logic and Scientific Method
London School of Economics and Political Science, United Kingdom
alexandru.marcoci@gmail.com

Institute for Logic, Language and Computation (ILLC)
Universiteit van Amsterdam, The Netherlands

Yann Pequignot

HEC, Research Group in Mathematical Logic
Université de Lausanne, Switzerland
yann.pequignot@unil.ch

Damien Servais

Institut supérieur de Philosophie
Université catholique de Louvain, Belgium
damien.servais@uclouvain.be

Sylvia Wenmackers

Department of Theoretical Philosophy
Rijksuniversiteit Groningen, The Netherlands
s.wenmackers@rug.nl

Vieri Benci

Dipartimento di Matematica Applicata
Università di Pisa, Italy
benci@dma.unipi.it

Leon Horsten

Department of Philosophy
University of Bristol, United Kingdom
Leon.Horsten@bristol.ac.uk

Stefan Wintein

Tilburg Center for Logic and Philosophy of Science (TiLPS)
Tilburg University, The Netherlands
s.wintein@uvt.nl

Editors' Preface

This proceedings volume bundles the contributed papers presented at PhDs in Logic III (sorted by last name). PhDs in Logic is an annual series of conferences/winter schools in the Low Countries. After successful editions in Ghent (2009) and Tilburg (2010) the third edition took place in Brussels on February 17–18, 2011. It consisted of four tutorials by established researchers and fourteen contributed talks by PhD students. The topics ranged from philosophical to mathematical logic. Over the two days the conference attracted about 40 logicians, mainly from Belgium and the Netherlands, but also from Germany, France, Switzerland, Denmark, and Slovenia.[1]

The four tutorials consisted of two one-hour sessions. Mai Gehrke (Nijmegen) gave a tutorial on duality theory and its many applications in logic (e.g. canonical extensions). Peter Koepke (Bonn) gave a tutorial on set theory, discussing techniques for getting large cardinals in inner models, and their connection with infinitary combinatorial principles. Eric Pacuit (Tilburg) gave an overview of contemporary work in epistemic logic, focusing on three pairs of concepts: single-agent/multi-agent, hard/soft information, and static/dynamic. Finally, Sonja Smets (Groningen) gave a tutorial on quantum logic, presenting both classical work and the new modal/epistemic approach she has recently been developing with Alexandru Baltag.

The contributed talks can be classified in three main groups: algebraic logic, set theory, and philosophical logic. There were three talks on algebraic logic. Dion Coumans talked about her application of duality theory to obtain a relational (Kripke) semantics for a fragment of linear logic. Sam van Gool discussed his ongoing work on a 'step-by-step' method to finitely approximate the Lindenbaum algebras of various modal logics. Yann Pequignot presented a duality perspective on the equational theory of regular languages.

Five of the contributed talks were on set theory. Raphaël Carroy showed how to use the backtrack game to decompose a Baire class one function into a collec-

---

[1] A report of this event was published in the online newsletters `thereasoner.org` (March 2011; vol. 5, nr. 3, pp. 40–41) and `loriweb.org` (post of February 25, 2011). This editorial preface is partly based on these reports.

tion of continuous functions, thus inducing a fine-grained hierarchy on the Baire class one functions. Kevin Fournier presented his research on the Wadge hierarchy and reducibility by continuous functions. Yurii Khomskii provided an overview of regularity properties of sets of real numbers.[2] Philipp Lücke presented his research on automorphism towers; this group-theoretical construction turns out to be highly dependent upon the model of set theory in which it is computed. Damien Servais, finally, left the familiar terrain of ZF: he discussed his work on Quine's alternative axiomatization NF (New Foundations).

The remaining six talks were on various topics in philosophical logic. Vincent Degauquier discussed his work on four (families of) bivalent logics: classical, paraconsistent, paracomplete, and positive. He introduced sequent calculi for these logics and compared their proof-theoretic strength. Tjerk Gauderis introduced his modal adaptive approach to abduction. His reasons for 'going modal' were mainly syntactical in nature; during discussion an intuitive epistemic interpretation was suggested as well. Jens Ulrik Hansen proposed to model the social-psychological notion of 'pluralistic ignorance' in dynamic epistemic logic (DEL), arguing that dynamic operators are necessary to capture the fragility of this notion. Alexandru Marcoci presented his new, DEL-based, solution to the surprise examination paradox. Sylvia Wenmackers presented an interrelated set of problems with Kolmogorov's axiomatization of probability theory, and proposed a new one, based on non-Archimedian probabilities (with applications to fair infinite lotteries). Stefan Wintein talked about his work on formal theories of truth, arguing that Kremer's modification of the Gupta-Belnap desideratum about vicious reference should be modified once again, using notions of (strong) assertibility.

We wish to thank the following institutions for their financial support: CNRL/NCNL (Centre National de Recherches de Logique/Nationaal Centrum voor Navorsingen in de Logica), the Doctoral School of Humanities of the Vrije Universiteit Brussel, the Center for Logic and Philosophy of Science (CLWF) of the Vrije Universiteit Brussel, and KVAB (Koninklijke Vlaamse Academie van België voor Wetenschappen en Kunsten). The KVAB is to be thanked not only for their generous financial support, but also for providing the beautiful and historical setting (the Academy Palace) in which the conference and winter school took place.

We wish to thank Margaux Smets for her practical assistance before and during the conference, and for her help with editing this volume.

We wish to thank the following people for carefully refereeing one or several abstracts: Patrick Allo, Max Creswell, Leon Horsten, Chris Impens, Thierry Libert, Benedikt Löwe, Erik Weber, and Andreas Weiermann. There was one more referee, namely Paul Gochet. Gochet was one of the great names in logic, in Belgium and far abroad. Sadly, he passed away on June 21, 2011. Several of the obituaries that have appeared since mention Gochet's never-ceasing activity and enthusiasm for all things logical. We wish to add one more piece of evidence for this claim. When we asked him in December 2010 to referee one of the abstracts he immediately agreed and he sent us a detailed report just a few days later. Additionally, he apologized for 'only' having read the abstract itself—later he told us that normally he also reads all (!) papers mentioned in the abstracts and papers that he reviews. (Unfortunately, because of his physical condition, Gochet was not able to attend the actual conference in

---

[2]Khomskii elected not to have his paper included in this volume.

February.)

PhDs in Logic III was an academically successful and socially vivid event, illustrating the variety and quality of contemporary research in logic in Belgium, the Netherlands, and far beyond. The conference series was continued in 2012 with a fourth installment in Ghent (April 12–13, 2012).

May 2012,

Jonas De Vuyst and Lorenz Demey

CHAPTER 1

Decomposing Baire Class One Functions

Raphaël Carroy

Abstract

The first Baire class has been widely studied, and a natural way to refine the Baire hierarchy has been defined first topologically in Hertling (1996), then game-theoretically in (Ros 2007). We give an overview of known results of the field, giving some new ones along the way. We begin with a presentation of those games, then proceed as in Hertling (1996) and Ros (2007) to define the discontinuity of a function. We give some results concerning the link between discontinuity and Wadge degree, and finally we give a new game-theoretical proof of the game characterization result for the first Baire class.

1 Introduction

Reduction games are now a classical tool in descriptive set theory as far as totally disconnected spaces are concerned. They were introduced by Wadge in his PhD thesis (1983) to define the Wadge hierarchy of Borel sets, then used again by Duparc (2001) to get the full analysis of the Wadge hierarchy for Borel sets of finite rank. Semmes (2007) also defined new reduction games to get extensions of the Jayne-Rogers theorem. Ros (2011) is a recent general presentation of reduction games.

Refining Van Wesep's backtrack game, Motto Ros in his PhD thesis (2007) found a game characterisation of a hierarchy of the first Baire class of functions introduced by Hertling and Weihrauch (1994) (see also Pauly 2010). We show that the games defined by Motto Ros are determined for a fixed function. We will then recall the definition of the hierarchy. Although Hertling and Weihrauch (1994) speak about the level of a function f and Ros (2007) about the rank of f, we refer to the place of f in the hierarchy as the *discontinuity* of f because we think this name is somehow more enlightening.

We then present a link between the discontinuities and the Wadge hierarchy of Δ^0_2-sets. Whereas Ros (2007) was computing the Wadge degree of inverse images of open sets considering a function of fixed countable discontinuity, here we proceed the other way around and we compute the discontinuity of the characteristic function of a set of fixed Wadge degree.

Jayne and Rogers (1982) showed that a function is Δ^0_2-measurable if and only if (iff) it is piecewise continuous, and Andretta (2006) showed that the backtrack game characterizes piecewise continuous functions. The reduction game for Δ^0_2-measurable functions is thus already known, nevertheless we give an alternative proof of this by showing directly that a function is Δ^0_2-measurable iff its discontinuity is countable.

Preliminaries. A *polish* space is a separable and completely metrizable topological space. Given A and B two polish spaces, a function $f : A \to B$ is *of Baire class one* if f is the pointwise limit of a sequence of continuous functions, or equivalently if the inverse image of a Σ^0_1 subset of B is Σ^0_2 in A. A closed set is called Π^0_1, open sets are Σ^0_1, a countable union of Π^0_1 is Σ^0_2, a countable intersection of Σ^0_1 is Π^0_2, and a set is Δ^0_2 iff it is both Σ^0_2 and Π^0_2. We say that $A \subseteq B$ is *nowhere dense* if $\bar{A}$, the closure of A, has empty interior. $A \subseteq B$ is *meager* if there exists a family $(A_i)_{i \in \mathbb{N}}$ of nowhere dense sets such that $A = \bigcup_{i \in \mathbb{N}} A_i$. The space B is *a Baire space* if every non-empty open set in B is non-meager. Recall that every polish space is Baire.

We denote: $\omega^{<\omega}$ (resp ω^ω, resp $\omega^{\leq\omega}$) the set of finite (resp infinite, resp any) sequences of integers, for $u \in \omega^{<\omega}$ and $v \in \omega^{\leq\omega}$, put $u \subset v$ if u is a strict subsequence of v, $[u] = \{x \in \mathcal{N} : u \subset x\}$, and ε for the empty sequence, call $u^\frown v$ the concatenation of u and v, and $lg(u)$ the *length* of u. If $u \neq \varepsilon$ then $lg(u) > 0$ and we will then write $d(u)$ for the last element of u, in other words $u(lg(u) - 1)$. A *tree* T on ω is a subset of $\omega^{<\omega}$ closed under subsequences. If $u \in T$ we say that u is a *node* of T. A node with no extension in T is a *terminal node* or a *leaf* of T. Two nodes u and v are *incompatible* if neither $u \subset v$ nor $v \subset u$, a set A of pairwise incompatible nodes is an *antichain*. Let $x \in \omega^\omega$, if for all $n \in \mathbb{N}$, $x|_n \in T$ we say that x is a *branch* of T, and we write $[T]$ for the set of all branches of T. If for all $t \in T$ there is an $s \in T$ s.t. $t \subset s$ then we say that T is *pruned*. Notice that a pruned tree has no leaf. Using the axiom of dependent choice, we have that every pruned tree has a branch. The set $[u]$ for $u \in \omega^{<\omega}$ is the set of branches of a pruned tree on ω.

Now $\{[u] : u \in \omega^{<\omega}\}$ is a basis for a topology on ω^ω, which is thus a topological space that we call the *Baire space*, denoted $\mathcal{N}$. The map $T \mapsto [T]$ is a bijection between pruned trees on ω and closed subsets of $\mathcal{N}$, so every basic open set of $\mathcal{N}$ is also closed, we say that $\mathcal{N}$ is *totally disconnected*. This space is canonical amongst totally disconnected polish spaces, because any such space is homeomorphic to a closed subset of the Baire space. In this paper we will consider partial functions $f : \mathcal{N} \to \mathcal{N}$ whose domain $\mathsf{dom}(f)$ is a closed subset of $\mathcal{N}$.

Finally, recall that there is no infinitely decreasing sequence of ordinals, so let s be a decreasing sequence of ordinals, there is an integer $l(s)$ such that the sequence $(s(n))_{n > l(s)}$ is constant.

2 The Games we Play

Definitions. Consider a function $f \colon \mathcal{N} \to \mathcal{N}$, an ordinal $\alpha < \omega_1$, and a word $w \in \omega^{<\omega}$.

We call $\mathbb{G}_\alpha(f, w)$ the following two-player infinite game with perfect information: at turn number $i \in \mathbb{N}$, player I picks up an integer n_i, player II picks up a couple $(u_i, \beta_i) \in \omega^{<\omega} \times (\alpha + 1)$. Let $x := (n_i)_{i \in \mathbb{N}}$ be the sequence of I's moves. Player II wins the game iff

either $w^{\frown} x \notin \mathsf{dom}(f)$ or both the following conditions are fulfilled:

1. the sequence $(\beta_i)_{i \in \mathbb{N}}$ is decreasing, and for all integers i, j such that $i < j$ we have that $\beta_i = \beta_j \Rightarrow u_i \subseteq u_j$, and if $y := \bigcup_{i > l(\beta)} u_i$ is the sequence that II has produced, then y is an element of $\mathcal{N}$.

2. $f(w^{\frown} x) = y$.

A *strategy* for player II in the game $\mathbb{G}_\alpha(f, w)$ is a function

$$
\begin{aligned}
\sigma \colon \quad \omega^{<\omega} &\to \quad \omega^{<\omega} \times (\alpha + 1) \\
u &\mapsto \quad (\sigma_1(u), \sigma_2(u)),
\end{aligned}
$$

such that (s.t.) the first winning condition for II is fulfilled when II is playing following σ. If the second winning condition is also fulfilled, we shall say that σ is a *winning strategy* (w.s.). For $u \in \omega^{<\omega}$, we write σ_u the strategy such that for any $v \in \omega^{<\omega}$, we have $\sigma_u(v) = \sigma(u^{\frown} v)$. If w is the empty word, we write $\mathbb{G}_\alpha(f)$ instead of $\mathbb{G}_\alpha(f, w)$.

Lemma 2.1. *Player II has a w.s. in $\mathbb{G}_\alpha(f, w)$ iff there is a Π^0_1-set $F \subseteq \mathcal{N}$ s.t.:*

1. *If $F \cap [w] \neq \emptyset$ then II has a w.s. in $\mathbb{G}_0(f|_F, w)$, and*

2. *for all $u \in \omega^{<\omega}$, if $w \subseteq u$ and if $[u] \cap F = \emptyset$ then there exists $\beta \in \alpha$ s.t. II has a w.s. in $\mathbb{G}_\beta(f, u)$.*

Proof. Suppose first that II has a w.s. σ in $\mathbb{G}_\alpha(f, w)$. Then $\{u \in \omega^{<\omega} : \sigma_2(u) = \alpha\}$ is a tree T on the alphabet $\mathbb{N}$, so the set $F = [T]$ of all infinite branches of T, possibly empty, is a closed subset of $\mathcal{N}$. As σ is a w.s., the second condition is fulfilled by definition of T. If F is empty, the first condition is always true. Otherwise $u \mapsto (\sigma_1(u), 0)$ is a w.s. for II in $\mathbb{G}_0(f|_F, w)$.

Suppose now $F \subseteq \mathcal{N}$ is a closed set fulfilling conditions 1 and 2. Let σ be II's w.s. in $\mathbb{G}_0(f|_F, w)$, and let $A \subset \omega^{<\omega}$ be an antichain s.t. $\bigcup_{u \in A}[u] = F^c \cap [w]$. For each $u \in A$, let $\beta_u \in \alpha$ be an ordinal and σ^u be II's w.s. in $\mathbb{G}_{\beta_u}(f, u)$. We now define a new strategy σ'. For any $v \in \omega^{<\omega}$,

$$
\sigma'(v) = \begin{cases} (\sigma_1(v), \alpha) & \text{if } [v] \cap F \neq \emptyset \\ \sigma^u(v) & \text{if there exists } u \in A \text{ s.t. } u \subseteq v. \end{cases}
$$

There are two possibilities when I plays along $x \in \mathcal{N}$. Either $x|_i \cap F \neq \emptyset$ for all $i \in \mathbb{N}$, then II will play following σ_1 and win because as F is closed, $x \in F$; or I will play a $u \in A$, and II will play following σ^u and win. Anyway, σ' is a w.s. $\qquad \square$

Definition. A *strategy for player I* in $\mathbb{G}_\alpha(f, w)$ is a function

$$\tau : \omega^{<\omega} \times \omega^{<\omega} \times (\alpha + 1) \to \mathbb{N}.$$

A strategy τ for I is *winning* if for any strategy σ for player II, the unique run of the game determined by the couple (τ, σ) is such that I wins. To be more accurate, given a couple (τ, σ) of strategies, define a sequence $x \in \mathcal{N}$ recursively: let n be an integer, then $x(n) = \tau(x|_n, \sigma(x|_n))$. Define then $y = \bigcup_{i>l(\sigma_2)} \sigma_1(x|_i)$. We say that (x, y) is *the run of the game determined by* (τ, σ).

Proposition 2.2. *If* $f : \mathcal{N} \to \mathcal{N}$ *is discontinuous, then I has a w.s. in* $\mathbb{G}_0(f)$.

Proof. The function f is discontinuous, so let $(x_n)_{n\in\mathbb{N}}$ be a sequence in $\mathcal{N}$ converging to $x \in \mathcal{N}$ and s.t. $\lim_{n\in\mathbb{N}} f(x_n) \neq f(x)$. Then let $N \in \mathbb{N}$ be s.t.

$$\forall n \in \mathbb{N} \, \exists i \geq n \, (f(x_i) \notin [f(x)|_N]).$$

Player II's moves are couples $(0, u)$, for $u \in \omega^{<\omega}$, so it suffices to describe the strategy for I without taking care of the ordinal part of II's moves, so let I play $x(i)$ at turn $i \in \mathbb{N}$. There are two possibilities:

1. II plays $y_j \in \omega^{<\omega}$ for some $j \in \mathbb{N}$ such that $f(x)|_N \nsubseteq y_j$, then $y_j \subset y = \bigcup y_n \neq f(x)$ and I wins.

2. for one $i \in \mathbb{N}$ II plays y_i such that $f(x)|_N \subseteq y_i$. Player I has just made the move $x(i)$. As $\lim_{n\in\mathbb{N}} x_n = x$, there is an integer m such that for all $n > m$, $x_n \in [x|_{i+1}]$, and there is $j \geq m$ such that $f(x_j) \notin [f(x)|_N]$. Then I plays $x_j(i + 1)$, and wins. □

Theorem 2.3. *For all* $(\alpha, u, f) \in \omega_1 \times \omega^{<\omega} \times \mathcal{N}^{\mathcal{N}}$, *the game* $\mathbb{G}_\alpha(f, u)$ *is determined.*

Proof: We proceed by induction on α.
Let us first consider the case $\alpha = 0$. Suppose $f : \mathcal{N} \to \mathcal{N}$ a continuous function. Consider the following function:

$$
\begin{aligned}
\sigma : \quad & \omega^{<\omega} && \to && \omega^{<\omega} \\
& s && \mapsto && u \text{ s.t. } \mathrm{lg}(u) = \max\{\mathrm{lg}(v) : v \in \omega^{\mathrm{lg}(s)} \wedge f([s]) \subseteq [v]\}.
\end{aligned}
$$

The function f being continuous, for all $x, y \in \mathcal{N}$, for all $N \in \mathbb{N}$ there is an $n \in \mathbb{N}$ such that $x|_n = y|_n$ implies $f(x)|_N = f(y)|_N$. In other words, σ is a strategy for II. Indeed, this strategy is winning. So if II has no w.s. then the function is discontinuous, so by Proposition 2.2, I has a w.s., and the game is determined.

Suppose now that for all $\beta < \alpha$, the game $\mathbb{G}_\beta(f, w)$ is determined. Let us prove now that if II has no w.s. in $\mathbb{G}_\alpha(f, w)$, then I has one.
By Lemma 2.1, and the induction hypothesis, for every closed set F,

1. either $F \cap [w] \neq \emptyset$ and I has a w.s. in $\mathbb{G}_0(f|_F, w)$,

2. or there exists $u \in \omega^{<\omega}$ s.t. $w \subseteq u$ and $[u] \cap F = \emptyset$ and for all $\beta < \alpha$, I has a w.s. in $\mathbb{G}_\beta(f, u)$.

Replacing F by $\emptyset$ in what precedes implies that $T = \{u \in \omega^{<\omega} : \forall \beta \in \alpha$ I has a w.s. in $\mathbb{G}_\beta(f, w^\frown u)\}$ is a non-empty tree. If T were well-founded, then by playing (ε, α), where ε is the empty word, until I plays out of T, II would have a w.s., so by contraposition $[T] \neq \emptyset$.

The set $[T]$ is non-empty closed, and does not verify point 2, so I has a w.s. in $\mathbb{G}_0(f|_{[T]}, w)$, that he can follow as long as II plays couples (s, α) for some $s \in \omega^{<\omega}$, and he wins if II does not change his moves. If II changes then he plays a couple (s, β), with $\beta \in \alpha$, which means, following the definition of T, that I has a w.s., so $\mathbb{G}_\alpha(f, w)$ is determined. □

3 The Discontinuity

We assume here that the reader is familiar with Borel derivatives and co-analytic ranks (see Kechris 1994, Ch. IV.34 or Moschovakis 2009, Ch. 4B). Define now, given a function $f \colon \mathcal{N} \to \mathcal{N}$ and F a closed subset of $\mathcal{N}$, the set

$$D(f, F) := F \setminus \bigcup \{[u] : u \in \omega^{<\omega} \text{ s.t. II has a w.s. in } \mathbb{G}_0(f|_F, u)\}.$$

This set is the closure of the set of discontinuity points of f in F so the map D is a Borel derivative, inducing a rank on F that we denote $|(f, F)|_D$, it is a countable ordinal. We then modify a bit this definition to get the discontinuity: we put $\text{disc}_f(F) = \sup\{\xi : D_\xi(f, F) \neq \emptyset\}$ if there is an α s.t. $D_\alpha(f, F) = \emptyset$ and put $\text{disc}_f(F) \geq \omega_1$ otherwise.

We also denote $\text{disc}_f(u) = \text{disc}_f([u])$, $\text{disc}(f) = \text{disc}_f(\mathcal{N})$, and $\text{disc}_f(x, F)$ for the discontinuity of x in F.

That finally leads us to define, given a function $f \colon \mathcal{N} \to \mathcal{N}$, the sets $D_{R\alpha}(f, F) = \{x \in \mathcal{N} : \text{disc}_f(x)R\alpha\}$, for $R \in \{<, >, =, \leq, \geq\}$. For short we put $\text{disc}_f(x) = \text{disc}_f(x, \mathcal{N})$ and $D_{R\alpha}(f) = D_{R\alpha}(f, \mathcal{N})$, for $R \in \{<, >, =, \leq, \geq\}$.

We have the following basic properties:

Proposition 3.1. *Given $f \colon \mathcal{N} \to \mathcal{N}$, and $F \subseteq \text{dom}(f)$ then*

1. *the sets $D_{>\alpha}(f, F)$ and $D_{\geq\alpha}(f, F)$ (resp. $D_{<\alpha}(f, F)$ and $D_{\leq\alpha}(f, F)$) are closed (resp open) in F,*

2. *If $\alpha < \omega_1$, and $x \in D_{=\alpha}(f)$, then for all $\beta < \alpha$ there is a sequence $(x_n)_{n\in\mathbb{N}} \subset D_{=\beta}(f)$ s.t. (x_n) converges to x,*

3. *For any function $f \colon \mathcal{N} \to \mathcal{N}$, the set $D_{\geq\omega_1}(f)$ is either empty or perfect,*

4. *for all $\alpha < \omega_1$, $f|_F$ is continuous on $D_{=\alpha}(f, F)$.*

Proof. Remark first that if $\text{disc}_f(u) < \omega_1$ then $\text{disc}_f(u) = \min\{\alpha \in \omega_1 :$ II has a w.s. in $\mathbb{G}_\alpha(f, u)\}$, so $\text{disc}_f(x, F) = \min\{\text{disc}_f([x|_n] \cap F) : n \in \mathbb{N}\}$. Then notice that, as D is a Borel derivative, we have for $u \in \omega^{<\omega}$ that $\text{disc}_f(u) = \sup\{\text{disc}_f(v) : u \subset v\}$, and if $\text{disc}_f(u) \in \omega_1$ we have furthermore that $\{\text{disc}_f(v) : u \subset v\}$ is an initial segment of ω_1.

1. We have that $D_{<\alpha}(f) = \bigcup_{\text{disc}_f(u)<\alpha}[u]$ and $D_{\geq\alpha}(f) = (D_{<\alpha}(f))^c$ (and the same for $\leq$ and $>$).

2. For all $\beta < \text{disc}(f)$, the closure by subsequences of $\{u \in \omega^{<\omega} : \text{disc}_f(u) = \beta+1\}$ is a pruned tree. Conclude by the remarks.

3. Just notice that on an isolated point of F, the function $f|_F$ is continuous.

4. By definition of $D_{=\alpha}(f)$, II has a w.s. in $\mathbb{G}_0(f|_{D_{=\alpha}(f)})$. □

Link with the Wadge hierarchy We assume here that the reader is familiar with classical results concerning the Wadge hierarchy. Those results can be found in Kechris (1994, Ch. II.21.E), Van Wesep (1978) and Duparc (2001). Now $\leq_W$ denotes the Wadge ordering on sets, and 1_X is the characteristic function of X, which means $1_X(a) = 1^\omega$ if $a \in X$, 0^ω otherwise.

Proposition 3.2. *For all sets A, B s.t. $\mathrm{disc}(1_A), \mathrm{disc}(1_B) < \omega_1$,*

1. *$A \leq_W B$ implies $\mathrm{disc}(1_A) \leq \mathrm{disc}(1_B)$,*

2. *$\mathrm{disc}(1_{B^c}) = \mathrm{disc}(1_B)$,*

3. *If $(B_i)_{i \in \mathbb{N}}$ is a family of sets s.t. $\mathrm{disc}(1_{B_i}) < \omega_1$ for $i \in \mathbb{N}$, then $A \equiv_W \bigsqcup_{i \in \mathbb{N}} B_i$ implies $\mathrm{disc}(1_A) = \sup\{\mathrm{disc}(1_{B_i}) : i \in \mathbb{N}\}$,*

4. *If $\mathrm{disc}(1_A) = \alpha$ then $\mathrm{disc}(1_{A+1}) = \alpha + 1$.*

Here $A + 1 = \{0^{2i}\frown x : i \in \mathbb{N}, x \in A\} \cup \{0^{2i+1}\frown x : i \in \mathbb{N}, x \in A^c\}$, and $\bigsqcup_{i \in \mathbb{N}} B_i$ is the disjoint union of all the B_i's.

Proof.

1. Suppose $A \leq_W B$. Let us now prove that for all $\alpha < \omega_1$, if II has a w.s. in $\mathbb{G}_\alpha(1_B)$ then he also has one in $\mathbb{G}_\alpha(1_A)$, then by definition of discontinuity we will be done.

   Now there is an increasing function $\sigma_0 \colon \omega^{<\omega} \to \omega^{<\omega}$ s.t. for all $x \in \mathcal{N}$, $1_A(x) = 1_B(\bigcup_{n \in \mathbb{N}} \sigma(x|_n))$. Let then σ be a w.s. for II in $\mathbb{G}_\alpha(1_B)$, $\sigma' = \sigma \circ \sigma_0$ is a w.s. for II in $\mathbb{G}_\alpha(1_A)$.

2. Just remark that if σ is a w.s. in $\mathbb{G}_\alpha(1_B)$, then σ' the strategy swapping the 0's and 1's is winning for II in $\mathbb{G}_\alpha(1_{B^c})$.

3. Put $\beta_i = \mathrm{disc}(1_{B_i})$ and $\beta = \sup\{\beta_i : i \in \mathbb{N}\}$. The union of the B_i's is disjoint so after finitely many rounds, I will play in one of them, II only has to wait for that round, then she has a w.s. in $\mathbb{G}_\beta(1_A)$ so $\mathrm{disc}(1_A) \leq \beta$. But for $\alpha \in \beta$, there is an integer i s.t. I has a w.s. in $\mathbb{G}_\alpha(1_{B_i})$ so he also has one in $\mathbb{G}_\alpha(1_A)$, hence $\beta = \mathrm{disc}(1_A)$.

4. Player II has a w.s. in $\mathbb{G}_{\alpha+1}(1_{A+1})$: as long as I plays 0, so does II. If I changes then II erases and apply the w.s. she has in $\mathbb{G}_\alpha(1_A)$ or in $\mathbb{G}_\alpha(1_{A^c})$. But I has a w.s. in $\mathbb{G}_\alpha(1_{A+1})$: playing 0 until II decides to play. If II plays in [1] then I goes on with 0, forcing II to erase or lose, but if II erases then I has a w.s. because $\mathrm{disc}(1_A) = \alpha$. If II plays in [0] then by playing either in A or in A^c I will force II to erase or lose, thus gaining a w.s. anyway. □

Let us now denote $d_W(A)$ the (coarse) Wadge degree of a Borel set A. Consider the following example:

$$f_0 : \quad \mathcal{N} \rightarrow \mathcal{N}$$
$$x \mapsto \begin{cases} (\frac{x(i)}{2})_{i \in \mathbb{N}} & \text{if } \forall i \ (x(i) \text{ is even}) \\ x & \text{otherwise.} \end{cases}$$

Here II has a w.s. in $\mathbb{G}_1(f_0)$ but f_0 is clearly not continuous, so $\mathsf{disc}(f_0) = 1$, yet if $F_0 = \{0^i\hat{\ }2^j\hat{\ }1^\omega : (i,j) \in \mathbb{N}^2, j > 0\} \cup \{0^i\hat{\ }2^\omega : i \in \mathbb{N}\} \cup \{0^\omega\}$, we have that F_0 is closed but $d_W(f^{-1}(F_0)) = 6$. Hence there is no obvious correspondence between discontinuity and the Wadge refinement of the Borel hierarchy of functions. However we can bound the Wadge degree of inverse images of open sets knowing the (countable) discontinuity of a function. Write $n/2$ for the quotient in the euclidian division of n by 2, and put $\eta(\alpha, n) = 0$ if $\alpha = 0$ or $n = 0$, 1 otherwise.

Corollary 3.3. *1. If A is a Δ_2^0 subset of $\mathcal{N}$, and n is the unique integer s.t. $d_W(A) = \alpha + n$, with $\alpha < \omega_1$ a limit ordinal, then $\mathsf{disc}(\mathbf{1}_A) = \alpha + (n/2) + \eta(\alpha, n)$,*

2. If $f : \mathcal{N} \rightarrow \mathcal{N}$ is a function s.t. $\mathsf{disc}(f) < \omega_1$, then the inverse image of any open set is Δ_2^0.

Proof.

1. By Proposition 3.2, if A is of a non-self-dual degree and B of the self-dual degree immediately above then $\mathsf{disc}(\mathbf{1}_A) = \mathsf{disc}(\mathbf{1}_B)$ but if B is of another degree then $\mathsf{disc}(\mathbf{1}_A) \neq \mathsf{disc}(\mathbf{1}_B)$, which gives us the result.

2. Let us first show, given two functions h and g, how to bound the discontinuity of $h \circ g$ using $\mathsf{disc}(h)$ and $\mathsf{disc}(g)$. Suppose that $\alpha = \mathsf{disc}(h)$, $\beta = \mathsf{disc}(g)$, $\sigma = (\sigma_0, \sigma_1)$ is a w.s. for II in $\mathbb{G}_\alpha(h)$, $\tau = (\tau_0, \tau_1)$ a w.s. for II in $\mathbb{G}_\beta(g)$.

   Fix Γ an isomorphism between the set $(\alpha+1)\times(\beta+1)$ ordered lexicographically and the ordinal $\delta = (\beta+1).\alpha + \beta$. Then

$$\rho : \quad \omega^{<\omega} \longrightarrow (\delta+1) \times \omega^{<\omega}$$
$$u \mapsto (\Gamma(\sigma_0 \circ \tau_1(u), \tau_0(u)), \sigma_1 \circ \tau_1(u))$$

   is a w.s. for player II in $\mathbb{G}_\delta(h \circ g)$.

   But now if $U \subseteq \mathcal{N}$ is any open set, we have that $\mathsf{disc}(\mathbf{1}_{f^{-1}(U)}) = \mathsf{disc}(\mathbf{1}_U \circ f) \leq \mathsf{disc}(f).2 + \mathsf{disc}(f)$, so we have our result. $\square$

Corollary 3.4. *If A is a Δ_2^0 subset of $\mathcal{N}$, then for all perfect sets F there are two open sets $A_1 \subseteq A$ and $A_2 \subseteq A^c$ such that $A_1 \cup A_2$ is open and dense in F.*

Proof. Let F be any perfect set in $\mathcal{N}$. The set A is Δ_2^0, so Corollary 3.3 implies $\mathsf{disc}(\mathbf{1}_A|_F) < \omega_1$. So $U = D_0(\mathbf{1}_A) \cap F$ is an open set of F on which $\mathbf{1}_A$ is continuous. This means $\mathbf{1}_A|[u]$ is constant equal either to 1^ω or 0^ω for all $[u]$ basic open in U, and so we are done because density is implied by Proposition 3.1. $\square$

4 Study of Baire Class One Functions

Definition. Considering a function $g : \mathcal{N} \to \mathcal{N}$, we define the set

$$C(g) := \{(u, v) \in \omega^{<\omega} \times \omega^{<\omega} : [u] \cap \mathrm{dom}(f) \neq \emptyset \wedge g([u]) \subseteq [v]\},$$

along with the following order: $(u, v) <_g (u', v')$ iff $v' \subset v$ and $u' \subset u$.

There is a one-to-one function from the set of all infinite decreasing sequences in $(C(g), <_g)$ onto the set of continuity points in the graph of g. Just consider for each such sequence $s = (s_0, s_1)$ the point $(x, y) = (\bigcup s_0, \bigcup s_1)$. So the continuity points of g are dense in $\mathrm{dom}(f)$ iff $(C(g), <_g)$ has no minimal element.

Theorem 4.1. *(Baire) A function* $f : \mathcal{N} \to \mathcal{N}$ *is Baire class one iff for any closed set* $A \subseteq \mathrm{dom}(f)$, *the continuity points of* $f|_A$ *are dense in* A.

Proof. ($\Rightarrow$): Suppose f is Baire class one. By Proposition 3.1, continuity points are dense in every closed subset of $D_{<\omega_1}(f)$. On the other hand, isolated points are dense in every countable closed set, so it suffices to show that if P is a perfect subset of $D_{\geq \omega_1}(f)$ and $g = f|_P$, $(C(g), <_g)$ has no minimal element.

Let $(u, v) \in C(g)$ and $w \supset v$, g is Baire class one so $g^{-1}([w])$ is Δ_2^0. If for all such w, $g^{-1}([w])$ has empty interior, then as all are Δ_2^0, complements $g^{-1}([w])^c$ contain a dense open set by corollary 3.4.

The set P having the Baire property induces that $\bigcap_{w \supset v} g^{-1}([w])^c$ contains a dense open set, so $g^{-1}([v])$ has empty interior. This is absurd because $(u, v) \in C(g)$, hence (u, v) is not minimal.

($\Leftarrow$): Suppose f is not Baire class one. Then there is a clopen subset A s.t. the inverse image of A is not Δ_2^0. So by Corollary 3.4 there is a perfect set $F \subseteq \mathcal{N}$ s.t. both $f^{-1}(A)$ and $f^{-1}(A)^c$ are dense in F. Suppose x is a continuity point of $f|_F$. Now there are sequences (x_n) and (y_n), both converging to x, s.t. for all $n \in \mathbb{N}$, $f(x_n) \in A$ and $f(y_n) \in A^c$. But, as x is a continuity point of $f|_F$ and A is clopen, we have that $f(x)$ is both in A and A^c, a contradiction. □

Definitions. Let $f : \mathcal{N} \to \mathcal{N}$ be any function, $\alpha \in \omega_1$.

A strategy σ for player II in $\mathbb{G}_\alpha(f)$ is *non-losing* if for all $u \in \omega^{<\omega}$ there exists $v \in \omega^{<\omega}$ s.t.

if $u \subseteq v$, $[\sigma_1(u)] \cap f([v]) = \emptyset$, and $\sigma_2(u) \neq 0$ then there is

$w \supseteq v$ s.t. $\sigma_2(w) \in \sigma_2(u)$ and $f([w]) \cap [\sigma_1(w)] \neq \emptyset$.

Intuitively if player II is 'outside' the image of what I did, and if she can still erase, then she does and she corrects.

A strategy σ for player II in $\mathbb{G}_\alpha(f)$ is *optimal* if it is non-losing, and for all u, v in $\omega^{<\omega}$,

1. Player II does not erase unless she is absolutely sure to be wrong-going, so if $u \subseteq v$ and $\sigma_2(v) \in \sigma_2(u)$ then $f([v]) \cap [\sigma_1(u)] = \emptyset$,

2. Whenever II has to take a guess, which happens every time I persists in playing somewhere discontinuous, then II bets on the discontinuity. Formally, if $f([u]) \subseteq [\sigma_1(u)]$, $u \subset v$, $\sigma_2(u) = \sigma_2(v)$ and if for all $w \supset \sigma_1(u)$, $(v, w) \notin C(f)$ then for all $w \supset v$, $(w, \sigma_1(v)) \notin C(f)$.

Theorem 4.2. *Consider a Baire class one function $f : \mathcal{N} \to \mathcal{N}$. There are two sequences of equivalences:*

1. *(a) The inverse image of some open set is Σ_2^0-complete.*

   (b) The discontinuity of f is at least ω_1. Then we write $\mathsf{disc}(f) = \omega_1$.

   (c) There is a perfect set P s.t. the discontinuity points of $f|_P$ are dense in P.

2. *(a) The inverse image of any open set is Δ_2^0.*

   (b) The discontinuity of f is strictly less than ω_1.

   (c) For any perfect set P, the continuity points of $f|_P$ contain a dense open set of P.

Proof. Notice on the first hand that the two sequences are dual, because as f is Baire class one, $(1a) \Leftrightarrow \neg(2a)$, $(1b) \Leftrightarrow \neg(2b)$ and $(1c) \Leftrightarrow \neg(2c)$ by definition.

On the second hand, $(2b) \Rightarrow (2c)$ by Proposition 3.1, and Proposition 3.3 implies that $(2b) \Rightarrow (2a)$.

Remark also that if $\mathsf{dom}(f)$ does not contain a perfect set then the result is true, once again because f is continue in any isolated point. This is why we can assume that $\mathsf{dom}(f) = \mathcal{N}$.

Assuming $(2c)$ and using the same construction we used in the proof of Proposition 3.3, we can find an ordinal $\alpha < \omega_1$ and an increasing sequence $\{U_\beta : \beta \in \alpha\}$ of open sets s.t. $\bigcup_{\beta < \alpha} U_\beta = \mathcal{N}$, U_β is dense in $F_\beta = (\bigcup_{\gamma < \beta} U_\gamma)^c$ and f is continuous on $U_\beta \cap F_\beta$. Then we can see inductively that for any $u \in \omega^{<\omega}$, $\mathsf{disc}_f(u) = \min\{\eta \leq \alpha : [u] \in U_\eta\}$, hence $(2b)$.

As a matter of fact, we show that $(1b) \Rightarrow (1a)$ and we will be done.

Suppose that I has a w.s. in $\mathbb{G}_\alpha(f)$, for all $\alpha < \omega_1$. We define a partial function $\tau : \omega^{<\omega} \times \omega^{<\omega} \times \omega_1$ s.t. for all $\alpha \in \omega_1$, the strategy $\tau_\alpha := \tau|_{\omega^{<\omega} \times \omega^{<\omega} \times (\alpha+1)}$ is a w.s. for I in $\mathbb{G}_\alpha(f)$.

We will define τ only on $\{u \in \omega^{<\omega} : [u] \cap D_{\geq \omega_1}(f) \neq \emptyset\} \times \omega^{<\omega} \times \omega_1$. It means we can suppose that at any round of the game, the current position $u \in \omega^{<\omega}$ of player I following the strategy τ, is s.t. $\mathsf{disc}_f(u) \geq \omega_1$.

The strategy τ_0 exists by Proposition 2.2 because f is discontinuous. Suppose now that τ_β is defined, for all $\beta < \alpha$. For all strategies σ for player II in $\mathbb{G}_\alpha(f)$, and all $u \in \omega^{<\omega}$ s.t. $\sigma_2(u) = \alpha$, fix $\tau(u, \sigma_1(u), \alpha) = \tau(u, \sigma_1(u), 0)$. So for all σ following which II never erases, I wins because τ_0 is a w.s. If now II erases, so for all strategies σ and all $u \in \omega^{<\omega}$ s.t. $\sigma_2(u) < \alpha$, there are two possibilities:

- There is a $\beta < \alpha, \beta \geq \sigma_2(u)$ and some strategy σ' for II in $\mathbb{G}_\beta(f)$ s.t. the position $(u, \sigma(u))$ is reached by (τ_β, σ'). Then $\tau(u, \sigma(u))$ is already defined, and σ_u is a strategy in $\mathbb{G}_{\sigma_2(u)}(f)$ against which τ_β is winning by induction hypothesis.

- For all $\beta < \alpha$ and all strategies σ' for II in $\mathbb{G}_\beta(f)$, the position $(u, \sigma(u))$ is not reached by (τ_β, σ'). But we supposed that $\mathsf{disc}_f(u) \geq \omega_1 > \sigma_2(u)$ so player I has a w.s. η in $\mathbb{G}_{\sigma_2(u)}(f, u)$, so we can fix $\tau(u, \sigma(u)) = \eta(\varepsilon, \varepsilon, \sigma_2(u))$ where ε is the empty word, and we are done.

Let S be the set of all optimal strategies σ for player II in $\mathbb{G}_\alpha(f)$ for some $\alpha < \omega_1$, we can then say that a position is reached by σ for some $\sigma \in S$ if it is reached by (τ_α, σ) for some $\alpha < \omega_1$. Consider now

$$P := \{x \in \mathcal{N} : \forall i \in \mathbb{N} \; \exists (\sigma, v, \beta) \in S \times \omega^{<\omega} \times \omega_1 \; (x|_i, v, \beta) \text{ is reached by } (\tau, \sigma)\},$$

notice now that if $h = f|_P$, then h is Baire class one, τ_α is a w.s. for I in $\mathbb{G}_\alpha(h)$ for all $\alpha < \omega_1$ so $\mathsf{disc}(h) \geq \omega_1$, and $\mathrm{D}_{\geq \omega_1}(h) = P$. In particular, P is perfect.

Consider now the set L of positions reached in P for which player II will lose for sure at the next turn, in other words, she cannot erase any more, and player I is about to push her in a losing position:

$$L = \{v \in \omega^{<\omega} : \exists \sigma \in S \; \exists u \in \omega^{<\omega} \; ((u, v, 0) \text{ is reached by } (\tau, \sigma)$$
$$\wedge \; [v] \cap h([u]) \neq \emptyset \wedge [v] \cap h([u \frown \tau(u, v, 0)]) = \emptyset)\}.$$

Let now W be the inverse image of the open set $U_L = \bigcup_{v \in L}[v]$. Take $u \in \omega^{<\omega}$ s.t. $[u] \cap P \neq \emptyset$, we can find σ optimal and (τ, σ) reaches $(u, v, 0)$ for some $v \in \omega^{<\omega}$ s.t. $h([u]) \cap [v] \neq \emptyset$: if it is not the case, u being reached anyway, two possibilities,

- either II is outside I's move, and we can suppose that two can erase one last time by definition of τ this will not change a thing,

- or II can still erase and we can suppose that she spoiled all erasing possibilities at the last move.

Moreover, as it is still optimal for II to wait as long as she wants after that point, at some point I has to play $u' \supset u$ s.t. for all $v' \supset v \; h([u']) \not\subseteq [v']$, so we can suppose that $h([u]) \not\subseteq [v]$. Hence there is a $u' \supset u$ s.t. $h([u']) \cap [v] = \emptyset$ so $v \in L$ and W is dense in P. The definition of L implies that W contains no continuity points so we finally have by Corollary 3.4 and Theorem 4.1 that W is Σ_2^0-complete. $\qquad \square$

Acknowledgments

This research was supported by the Swiss National Science Foundation (no. 200021–116508). I would like to thank Jacques Duparc, Olivier Finkel, Kevin Fournier, Yann Pequignot and Marcin Sabok for their careful readings, remarks and corrections.

References

Andretta, A. (2006). "More on Wadge Determinacy". In: *Annals of Pure and Applied Logic* 144, pp. 2–32.

Duparc, J. (2001). "Wadge Hierarchy and Veblen Hierarchy Part I: Borel Sets of Finite Rank". In: *Journal of Symbolic Logic* 66, pp. 56–86.

Hertling, P. (1996). *Unstetigkeitsgrade von Funktionen in der effektiven Analysis*. Hagen: Fernuniversität Hagen, Fachbereich Informatik.

Hertling, P. and K. Weihrauch (1994). *On the Topological Classification of Degeneracies*. Hagen: Fernuniversität Hagen, Fachbereich Informatik.

Jayne, J. E. and C. A. Rogers (1982). "First Level Borel Functions and Isomorphisms". In: *Journal de Mathématiques Pures et Appliquées. Neuvième Série* 61, pp. 177–205.

Kechris, A. S. (1994). *Classical Descriptive Set Theory*. New York, NY: Springer Verlag.

Moschovakis, Y. N. (2009). *Descriptive Set Theory*. Providence, RI: American Mathematical Society.

Pauly, A. (2010). "On the (Semi)lattices Induced by Continuous Reducibilities". In: *Mathematical Logic Quarterly* 56, pp. 488–502.

Ros, L. M. (2007). "General Reducibilities for Sets of Reals". PhD thesis. Polytechnic of Turin.

— (2011). "Game Representations of Classes of Piecewise Definable Functions". In: *Mathematical Logic Quarterly* 57, pp. 95–112.

Semmes, B. (2007). "Multitape Games". In: *Interactive Logic: Selected Papers from the 7th Augustus de Morgan Workshop, London*. Ed. by J. van Benthem, D. Gabbay, and B. Löwe. Amsterdam: Amsterdam University Press, pp. 195–207.

Van Wesep, R. (1978). "Wadge Degrees and Descriptive Set Theory". In: *Cabal Seminar 76–77*. Ed. by A. S. Kechris and Y. N. Moschovakis. Lecture Notes in Mathematics 689. New York: Springer, pp. 151–170.

Wadge, W. (1983). "Reducibility and Determinateness on the Baire Space". PhD thesis. University of California, Berkeley.

CHAPTER 2

Relational Semantics for a Fragment of Linear Logic

Dion Coumans

Abstract

Relational semantics, given by Kripke frames, play an essential role in the study of modal and intuitionistic logic. In Dunn, Gehrke, and Palmigiano (2005) it is shown that the theory of relational semantics is also available in the more general setting of substructural logic, at least in an algebraic guise. Building on these ideas, in Gehrke (2006) a type of frames is described which generalize Kripke frames and provide semantics for substructural logics in a purely relational form.

We will extend the work in Dunn, Gehrke, and Palmigiano (2005) and Gehrke (2006) and use their approach to obtain relational semantics for multiplicative additive linear logic. Hereby we illustrate the strength of using canonical extensions to retrieve relational semantics: it allows a modular and uniform treatment of additional operations and axioms.

Traditionally, so-called phase spaces are used to describe semantics for linear logic (Girard 1987). These have the drawback that, contrary to our approach, they do not allow a modular treatment of additional axioms. However, the two approaches are related, as we will explain.[1]

1 Introduction

Relational semantics, given by Kripke frames, play an essential role in the study of modal and intuitionistic logic (Blackburn, de Rijke, and Venema 2001). They provide an intuitive interpretation of the logic and a means to obtain information about it.

---

[1] An extended version of this chapter will appear in the special issue 'Logic, categories, semantics' of the Journal of Applied Logic (Coumans, Gehrke, and Rooijen 2012).

The possibility of applying semantical techniques to obtain information about a logic motivates the search for relational semantics in a more general setting.

Many logics are closely related to corresponding classes of algebraic structures which provide *algebraic semantics* for the logics. The algebras associated to classical modal logic are Boolean algebras with an additional operator (BAOs). Kripke frames arise naturally from the duality theory for these structures in the following way. Boolean algebras are dually equivalent to Stone spaces (Stone 1936). A modal operator on Boolean algebras translates to a binary relation with certain topological properties on the corresponding dual spaces, hence giving rise to so-called descriptive general frames. Forgetting the topology yields Kripke frames, which are in a discrete duality with complex modal algebras, *i.e.*, modal algebras whose underlying Boolean algebra is a powerset algebra. This may be depicted as follows:

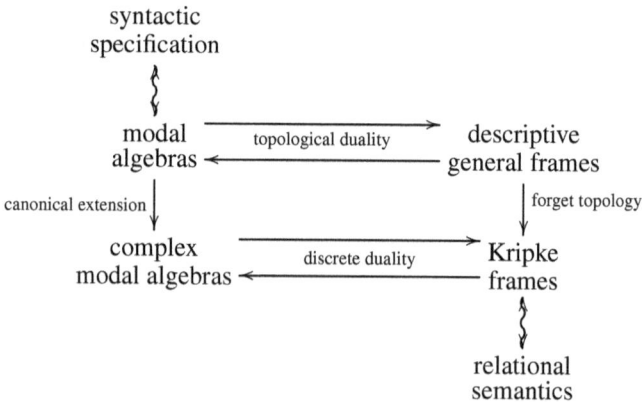

Hence, one may retrieve relational semantics for modal logic by first moving horizontally using the duality and thereafter going down by forgetting the topology.

Many other interesting logics, including substructural logics, however, have algebraic semantics which are not based on distributive lattices and for these duality theory is vastly more complicated or even non-existent. Luckily, the picture above also indicates an alternative route to obtain relational semantics: going down first and thereafter going right. The (left) downward mapping is given by taking the *canonical extension* of a BAO. Canonical extensions were introduced in the 1950s by Jónsson and Tarski exactly for BAOs (1951; 1952). Thereafter their ideas have been developed further, which has led to a smooth theory of canonical extensions applicable in a broad setting (Gehrke and Harding 2001; Gehrke and Jónsson 1994). In Dunn, Gehrke, and Palmigiano (2005) canonical extensions of partially ordered algebras are defined to obtain relational semantics for the fusion-implication fragment of various substructural logics. Their approach is purely algebraic. In Gehrke (2006) this work is translated to the setting of possible world semantics. A class of frames (RS-frames) is described which generalize Kripke frames and provide semantics for substructural logics in a purely relational form. This is summarized in the following picture:

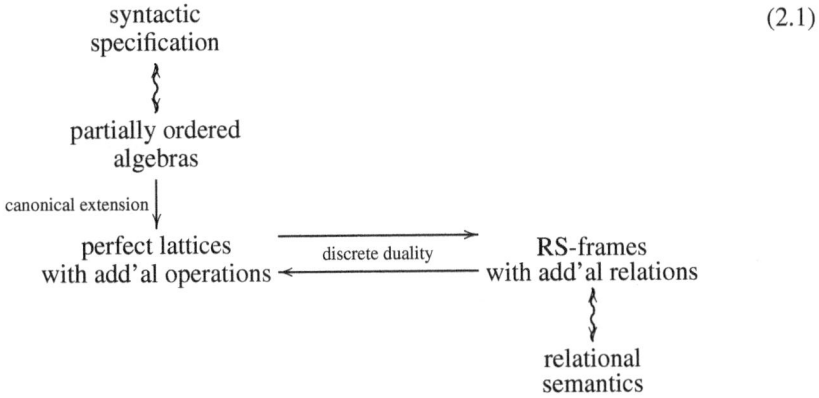

$$(2.1)$$

syntactic
specification

$\updownarrow$

partially ordered
algebras

canonical extension $\downarrow$

perfect lattices $\xrightarrow{\hspace{1cm}\text{discrete duality}\hspace{1cm}}$ RS-frames
with add'al operations $\longleftarrow$ with add'al relations

$\updownarrow$

relational
semantics

A well-know substructural logic is linear logic. Linear logic was introduced by Jean-Yves Girard (1987). Formulas in linear logic represent resources which may be used exactly once. Proof-theoretically this is witnessed by the fact that the structural rules contraction and weakening are not admissible in general. However, these structural rules are allowed in a controlled way by means of a new modality, the exponential !, which expresses the case of unlimited availability of a specific resource. Traditionally, phase spaces are used as semantics for linear logic. These have the drawback that, contrary to the approach described above, they do not allow a modular treatment of additional operations and axioms.

In this paper we discuss part of a joint project with Mai Gehrke and Lorijn van Rooijen on developing relational semantics for full linear logic. We focus on multiplicative additive linear logic (MALL), the fragment of linear logic that leaves the exponential out of consideration, and describe how to obtain relational semantics for MALL. Thereby we illustrate that canonical extensions allow a modular and uniform treatment of additional operations and axioms, which distinguishes our work from earlier derivations of Kripke-style semantics for linear logic (Allwein and Dunn 1993).

First, we discuss the general method of obtaining relational semantics for substructural logics using canonical extensions, essentially by explaining how to move 'down-right' in the picture above (Section 2) and by indicating how to show that this indeed yields complete relational semantics (Section 3). We focus on the parts of this general theory that are important for the remainder of our paper and refer the reader to Dunn, Gehrke, and Palmigiano (2005) and Gehrke (2006) for more details. In Section 4 this method is applied to obtain relational semantics for MALL. Finally, in Section 5 we discuss how our results relate to phase spaces.

2 Duality between Perfect Lattices and RS-Polarities

Algebraic semantics for substructural logics are given by partially ordered sets (posets) with additional operations on them (*partially ordered algebras*). Hence, the first step in obtaining relational semantics for substructural logics using the method depicted in Figure 2.1 is to define canonical extensions of posets. This is worked out in Section 2 of Dunn, Gehrke, and Palmigiano (2005) where one can find a careful and clear explanation of this theory. The structures arising as canonical extensions of posets are perfect lattices.

Definition 2.1. *A* perfect lattice *is a complete lattice that is both join-generated by its completely join-irreducible elements and meet-generated by its completely meet-irreducible elements.*

To move horizontally in Figure 2.1 one should identify relational structures that are in a duality with perfect lattices. In Gehrke (2006) a class of (two-sorted) frames fulfilling this requirement is described. We briefly discuss this duality.

Definition 2.2. *A (two-sorted) frame is a triple* $F = (X, Y, \leqslant)$ *where X and Y are sets and* $\leqslant \subseteq X \times Y$ *is a relation from X to Y.*

A frame gives rise to a Galois connection between $\wp(X)$ and $\wp(Y)$:

$$(\ )^u \ : \ \begin{array}{lcl} \wp(X) & \to & \wp(Y) \\ A & \mapsto & \{y \in Y \,|\, \forall x.\, x \in A \Rightarrow x \leqslant y\}, \end{array}$$

$$(\ )^l \ : \ \begin{array}{lcl} \wp(Y) & \to & \wp(X) \\ B & \mapsto & \{x \in X \,|\, \forall y.\, y \in B \Rightarrow x \leqslant y\}. \end{array}$$

The complete lattice of Galois-closed subsets of X is given by

$$\mathcal{G}(F) = \{A \subseteq X \,|\, (A^u)^l = A\},$$

which is a perfect lattice.

Conversely, for every perfect lattice **L**, we define a frame $\mathcal{F}(\mathbf{L})$ by $X = \mathcal{J}^\infty(\mathbf{L})$, $Y = \mathcal{M}^\infty(\mathbf{L})$ and, for all $x \in X, y \in Y$,

$$x \leqslant y \quad \Leftrightarrow \quad x \leq_\mathbf{L} y.$$

This frame is *separating*, i.e., the following two conditions hold:

1. $\forall x_1, x_2 \in X \ (x_1 \neq x_2 \Rightarrow \{x_1\}^u \neq \{x_2\}^u)$;
2. $\forall y_1, y_2 \in Y \ (y_1 \neq y_2 \Rightarrow \{y_1\}^l \neq \{y_2\}^l)$.

Furthermore it is *reduced*, i.e., the following two conditions hold:

1. $\forall x \in X \exists y \in Y \ (x \not\leqslant y \text{ and } \forall x' \in X \ [\{x'\}^u \supset \{x\}^u \Rightarrow x' \leqslant y])$,
2. $\forall y \in Y \exists x \in X \ (y \not\geqslant x \text{ and } \forall y' \in Y \ [\{y'\}^l \subset \{y\}^l \Rightarrow y' \geqslant x])$.

A frame that is both separating and reduced is called an *RS-frame*. The separating property implies that the maps

$$\begin{array}{lcl} X & \to & \mathcal{G}(F) \\ x & \mapsto & (\{x\}^u)^l \end{array} \qquad \begin{array}{lcl} Y & \to & \mathcal{G}(F) \\ y & \mapsto & \{y\}^l \end{array}$$

are injective. Therefore we may think of X and Y as subsets of $\mathcal{G}(F)$ and we will write x both for the element of X and for the corresponding element $\{x\}^{ul}$ of $\mathcal{G}(F)$ (and similarly for elements of Y). For an S-frame, being reduced exactly means that all elements of X are completely join-irreducible in $\mathcal{G}(F)$ and the elements of Y are completely meet-irreducible in $\mathcal{G}(F)$.

An RS-frame morphism $F_1 = (X_1, Y_1, \leq) \rightarrow (X_2, Y_2, \leq) = F_2$ is a pair of relations $S_1 \subseteq Y_1 \times X_2, S_2 \subseteq X_1 \times Y_2$ satisfying some conditions. These conditions ensure that the pair of relations gives rise to a complete lattice homomorphism $\mathcal{G}(S_1, S_2) \colon \mathcal{G}(F_2) \rightarrow \mathcal{G}(F_1)$. Conversely, for each complete lattice homomorphism $f \colon \mathbf{L_1} \rightarrow \mathbf{L_2}$ between perfect lattices, one may define an RS-frame morphism $\mathcal{F}(f) \colon \mathcal{F}(\mathbf{L_2}) \rightarrow \mathcal{F}(\mathbf{L_1})$.

Proposition 2.3. *The mappings $\mathcal{F}$ and $\mathcal{G}$ form a duality between the category of perfect lattices and the category of RS-frames.*

For further details and a proof of the above proposition, the reader is referred to Gehrke (2006).

3 Relational Semantics via Canonical Extension

We will now extend and apply the basic theory of the previous section to describe the general method for obtaining relational semantics for substructural logics.

The basic substructural logic we consider is non-associative Lambek calculus (NLC). Its signature consists of three binary operations $\otimes$, $\rightarrow$, $\leftarrow$. The axioms of NLC state that the implications $\rightarrow$ and $\leftarrow$ are residuals of the fusion $\otimes$. Algebraic semantics for this logic are given by residuated algebras.

Definition 3.1. *A residuated algebra is a structure $(\mathbf{P}, \otimes, \rightarrow, \leftarrow)$, where $\mathbf{P}$ is a partially ordered set and, for all $x, y, z \in \mathbf{P}$,*

$$x \otimes y \leq z \quad \Leftrightarrow \quad y \leq x \rightarrow z$$
$$\Leftrightarrow \quad x \leq z \leftarrow y.$$

A residuated algebra is called perfect *if its underlying poset is a perfect lattice.*

For a perfect residuated algebra, the underlying perfect lattice $\mathbf{L}$ corresponds dually to the RS-frame $\mathcal{F}(\mathbf{L}) = (\mathcal{J}^\infty(\mathbf{L}), \mathcal{M}^\infty(\mathbf{L}), \leq_{\mathbf{L}})$, as explained in Section 2. The action of the fusion (and thereby of its residuals) may be encoded on this dual frame as follows. First note that, as the fusion is residuated, it is completely join-preserving in both coordinates. Therefore, its action is completely determined by its action on pairs from $J^\infty(\mathbf{L}) \times J^\infty(\mathbf{L})$. Define a relation $R_\otimes \subseteq X \times X \times Y$ by

$$R_\otimes(x_1, x_2, y) \quad \Leftrightarrow \quad x_1 \otimes x_2 \leq y.$$

The relation $R_\otimes$ is *compatible*, that is, for all $x_1, x_2 \in X$, $y \in Y$, the sets

$$R_\otimes[x_1, x_2, \_] \qquad R_\otimes[x_1, \_, y] \qquad R_\otimes[\_, x_2, y]$$

are Galois-closed.[2]

Definition 3.2. *A structure $F = (X, Y, \preccurlyeq, R)$, where $(X, Y, \preccurlyeq)$ is an RS-frame and $R \subseteq X \times X \times Y$ is a compatible relation, is called a* relational RS-frame.

---

[2] We may also witness the fusion $\otimes$ dually by the relation $R_\downarrow \subseteq X^3$ defined by $R_\downarrow(x_1, x_2, x_3) \Leftrightarrow x_3 \leq x_1 \otimes x_2$. In that case, however, the conditions stating that R arises from a fusion are less natural.

Conversely, for an RS-frame $F = (X, Y, \leqslant)$, a relation $R \subseteq X \times X \times Y$ gives rise to a fusion $\otimes_R$ on $\mathcal{G}(F)$, by defining

$$x_1 \otimes_R x_2 = \bigwedge\{y \in Y \mid R(x_1, x_2, y)\} \qquad \text{for all } x_1, x_2 \in X,$$
$$w_1 \otimes_R w_2 = \bigvee\{x_1 \otimes_R x_2 \mid x_1 \leq u_1, x_2 \leq u_2\} \quad \text{for all } w_1, w_2 \in \mathcal{G}(F).$$

This operation is completely join-preserving in both coordinates and therefore it is residuated, with residuals $\rightarrow_R$ and $\leftarrow_R$.

For any residuated fusion operation $\otimes$ on a perfect lattice, $\otimes_{R_\otimes} = \otimes$ and, for any compatible relation R on an RS-frame, $R_{\otimes_R} = R$.

Proposition 3.3 (Proposition 6.6 in Dunn, Gehrke, and Palmigiano 2005). *The above defined maps $(\mathbf{L}, \otimes, \rightarrow, \leftarrow) \mapsto (\mathcal{F}(L), R_\otimes)$ and $(X, Y, \leqslant, R) \mapsto (\mathcal{G}(X, Y, \leqslant), \otimes_R, \rightarrow_R, \leftarrow_R)$ yield a duality between perfect residuated algebras and relational RS-frames.*[3]

In the remainder we will also denote the extended mappings of the above proposition (and any further generalizations) by $\mathcal{F}$ and $\mathcal{G}$.

One may define a satisfaction relation $\Vdash$ on relational frames, such that, for all frames F, for all formulas ϕ, ψ in NLC,

$$\phi \Vdash \psi \text{ holds in } F \qquad \Leftrightarrow \qquad \phi \leq \psi \text{ holds in } \mathcal{G}(F). \tag{2.2}$$

This is explained in detail in Section 4 of Gehrke (2006).

For a residuated algebra $\mathbf{P}$, the σ-extension of the fusion on $\mathbf{P}$, $\otimes^\sigma : \mathbf{P}^\delta \times \mathbf{P}^\delta \rightarrow \mathbf{P}^\delta$, is a residuated operator on the canonical extension $\mathbf{P}^\delta$ (Corollary 3.7 of Dunn, Gehrke, and Palmigiano 2005). This completes the description of the walk through Figure 2.1 for NLC: we start with a residuated algebra $\mathbf{P}$, its canonical extension is a perfect residuated algebra $\mathbf{P}^\delta$, which gives rise to a relational frame via the mapping $\mathcal{F}$.

We are now ready to describe our method for obtaining relational semantics for a substructural logic. Let $\mathcal{E}$ be a collection of inequalities axiomatizing a logic $\mathcal{L}_\mathcal{E}$ in the connectives $\otimes, \rightarrow, \leftarrow$, extending NLC. The collection $\mathcal{A}lg_\mathcal{E}$ of residuated algebras satisfying the inequalities in $\mathcal{E}$ provides complete algebraic semantics for $\mathcal{L}_\mathcal{E}$, in the sense that, for all formulas ϕ, ψ,

$$\phi \vdash \psi \text{ is derivable in } \mathcal{L}_\mathcal{E} \text{ iff } \phi \leq \psi \text{ holds in all residuated algebras in } \mathcal{A}lg_\mathcal{E}.$$

Our aim is to describe a collection of relational frames $\mathcal{K}$ which provides complete relational semantics for $\mathcal{L}_\mathcal{E}$. We define, for a collection of relational frames $\mathcal{K}$,

$$\mathcal{K}^+ = \{\mathcal{G}(F) \mid F \in \mathcal{K}\}.$$

By (2.2), $\mathcal{K}$ provides complete relational semantics for $\mathcal{L}_\mathcal{E}$ iff $\mathcal{L}_\mathcal{E} = \mathcal{E}qThr(\mathcal{K}^+)$, where $\mathcal{E}qThr(\mathcal{K}^+)$ is the equational theory of $\mathcal{K}^+$, *i.e.* the collection of inequalities that hold in all algebras in $\mathcal{K}^+$.

To obtain complete relational semantics for $\mathcal{L}_\mathcal{E}$ it suffices to obtain:

1. Canonicity: show that $\mathcal{A}lg_\mathcal{E}$ is closed under canonical extension, that is, show that, for all $\mathbf{P} \in \mathcal{A}lg_\mathcal{E}$, $\mathbf{P}^\delta \in \mathcal{A}lg_\mathcal{E}$.

---

[3]Note that we have not spelled out which morphisms we consider in both categories. The reader interested in more details is referred to Dunn, Gehrke, and Palmigiano (2005).

2. Correspondence: give necessary and sufficient conditions on relational frames F to ensure that $\mathcal{G}(F) \in \mathcal{A}lg_{\mathcal{E}}$.

Proposition 3.4. *If $\mathcal{A}lg_{\mathcal{E}}$ is closed under canonical extension, then $\mathcal{E}qThr(\mathcal{A}lg_{\mathcal{E}}) = \mathcal{E}qThr(\mathcal{A}lg_{\mathcal{E}}^{\delta})$, where $\mathcal{A}lg_{\mathcal{E}}^{\delta} = \{\mathbf{P}^{\delta} \mid \mathbf{P} \in \mathcal{A}lg_{\mathcal{E}}\}$.*

Proof. As, by assumption, $\mathcal{A}lg_{\mathcal{E}}^{\delta} \subseteq \mathcal{A}lg_{\mathcal{E}}$, clearly $\mathcal{E}qThr(\mathcal{A}lg_{\mathcal{E}}) \subseteq \mathcal{E}qThr(\mathcal{A}lg_{\mathcal{E}}^{\delta})$. For the converse, suppose $\phi \leq \psi$ holds in $\mathcal{A}lg_{\mathcal{E}}^{\delta}$ and $\mathbf{P} \in \mathcal{A}lg_{\mathcal{E}}$. As $\mathbf{P}$ embeds in its canonical extension $\mathbf{P}^{\delta}$ and $\mathbf{P}^{\delta} \in \mathcal{A}lg_{\mathcal{E}}^{\delta}$, $\phi \leq \psi$ holds in $\mathbf{P}$. □

If $\mathcal{A}lg_{\mathcal{E}}$ is closed under canonical extension we say the collection of axioms $\mathcal{E}$ is *canonical*. It follows from the above proposition that in this case the collection

$$\mathcal{K} = \{F \mid \mathcal{G}(F) \in \mathcal{A}lg_{\mathcal{E}}\}$$

provides complete relational semantics for $\mathcal{L}_{\mathcal{E}}$ (note that $\mathcal{A}lg_{\mathcal{E}}^{\delta} \subseteq \mathcal{K}^{+} \subseteq \mathcal{A}lg_{\mathcal{E}}$). In case the axioms in $\mathcal{E}$ are 'sufficiently simple' one may obtain, in a mechanical way, first-order conditions on relational frames F that are necessary and sufficient to ensure $\mathcal{G}(F) \in \mathcal{A}lg_{\mathcal{E}}$. Many well-known logics may be axiomatized by canonical and 'sufficiently simple' axioms, whence the above described procedure may be applied to obtain complete relational semantics. In Dunn, Gehrke, and Palmigiano (2005) this is worked out for the fusion-implication fragment of Lambeck calculus, linear logic, relevance logic, BCK logic and intuitionistic logic.

In case the logic is equipped with additional function symbols an extension of the above method may be applied. Algebraic semantics are given by residuated algebras equipped with additional operations (corresponding to the additional function symbols). To obtain relational semantics for the logic one has to give a description of these additional operations on relational frames. In the next section we will illustrate this procedure by deriving relational semantics for multiplicative additive linear logic.

4 Relational Semantics for MALL

To derive relational semantics for multiplicative additive linear logic (MALL), we start by describing its algebraic semantics. These are given by classical linear algebras, which are extensions of the residuated algebras studied in the previous section.

Definition 4.1. *A classical linear algebra (CL-algebra) is a structure $(\mathbf{L}, \otimes, \rightarrow, \leftarrow, 1, 0)$, where*

1. *$(\mathbf{L}, \otimes, \rightarrow, \leftarrow)$ is a residuated algebra;*

2. *the fusion $\otimes$ is associative and commutative and has a unit 1;*

3. *$\mathbf{L}$ is a bounded lattice;*

4. *for all $a \in \mathbf{L}$, $(a \rightarrow 0) \rightarrow 0 = a$.*

In linear logic, the meet operation is denoted by & (with unit $\top$), the join by $\oplus$ (with unit 0), the implication by $\multimap$ and our constant 0 is denoted by $\bot$. However, as we will refer to the literature from lattice theory we will stick to the usual lattice theoretic notation and denote meet by $\wedge$ (with unit $\top$) and join by $\vee$ (with unit $\bot$). For further details on CL-algebras the reader is referred to Troelstra (1992), which uses a notation similar to ours.

We will denote $x \to 0$ by $x^\perp$ and call this operation *linear negation*. Implication sends joins in the first coordinate to meets, hence $(\_)^\perp$ sends joins to meets. As $(\_)^\perp$ is a bijection, it follows that it is a (bijective) lattice homomorphism $\mathbf{L} \to \mathbf{L}^\partial$, where $\mathbf{L}^\partial$ is the lattice obtained by reversing the order in $\mathbf{L}$.

The first step in obtaining relational semantics for MALL is checking canonicity, *i.e.*, ensuring that the class **CL** of CL-algebras is closed under canonical extension.

Proposition 4.2. *The class* **CL** *is closed under canonical extension.*

Proof. Let $\mathbf{L}$ be a CL-algebra and let $\mathbf{L}^\delta$ be its canonical extension. In Dunn, Gehrke, and Palmigiano (2005) it is shown that $\mathbf{L}^\delta$ is a perfect residuated algebra. Hence, in particular, it is a bounded lattice. Furthermore, it is shown that, if $\otimes$ is associative (resp. commutative), then so is its extension $\otimes^\sigma$.

It is left to show that, for all $w \in \mathbf{L}^\delta$, $(w^{\perp^\delta})^{\perp^\delta} = w$. This may be derived from the results in Almeida (2009), but we choose to give a direct proof here to illustrate the methods of the theory of canonical extensions. As $(\_)^\perp$ is a lattice homomorphism $\mathbf{L} \to \mathbf{L}^\partial$, its extension is a complete lattice homomorphism $\mathbf{L}^\delta \to (\mathbf{L}^\partial)^\delta = (\mathbf{L}^\delta)^\partial$. Every element of the canonical extension may be written as a join of meets of elements of the original lattice. We write $K(\mathbf{L}^\delta)$ for the elements of $\mathbf{L}^\delta$ that may be obtained as a meet of elements of $\mathbf{L}$. For $w \in \mathbf{L}^\delta$,

$$
\begin{aligned}
w &= \bigvee \{x \in K(\mathbf{L}^\delta) \mid x \le w\} \\
&= \bigvee \{\bigwedge \{a \in \mathbf{L} \mid x \le a\} \mid x \in K(\mathbf{L}^\delta),\, x \le w\}.
\end{aligned}
$$

Hence, for $w \in \mathbf{L}^\delta$,

$$
\begin{aligned}
(w^{\perp^\delta})^{\perp^\delta} &= ((\bigvee \{\bigwedge \{a \in \mathbf{L} \mid x \le a\} \mid x \in K(\mathbf{L}^\delta),\, x \le w\})^{\perp^\delta})^{\perp^\delta} \\
&= (\bigwedge \{\bigvee \{a^\perp \mid a \in \mathbf{L},\, x \le a\} \mid x \in K(\mathbf{L}^\delta),\, x \le w\})^{\perp^\delta} \\
&= \bigvee \{\bigwedge \{(a^\perp)^\perp \mid a \in \mathbf{L},\, x \le a\} \mid x \in K(\mathbf{L}^\delta),\, x \le w\} \\
&= \bigvee \{\bigwedge \{a \in \mathbf{L} \mid x \le a\} \mid x \in K(\mathbf{L}^\delta),\, x \le w\} \\
&= w,
\end{aligned}
$$

which proves the claim. $\square$

To describe the constants 1 and 0 dually, we have to extend the relational frames with two Galois-closed subsets, $U \subseteq X$ and $Z \subseteq Y$. Starting from a perfect CL-algebra $\mathbf{L}$, these sets are given by

$$
U = \{x \in \mathcal{J}^\infty(\mathbf{L}) \mid x \le 1\} \quad \text{and} \quad Z = \{y \in \mathcal{M}^\infty(\mathbf{L}) \mid 0 \le y\}.[4]
$$

Our next step is to characterize the collection of frames $F = (X, Y, \le, R, U, Z)$ satisfying $\mathcal{G}(F) \in \mathbf{CL}$ (correspondence). In the remainder of this section, we assume that any element named x (resp. y) with any super- or subscript comes from X (resp. Y).

---

[4]We could also have described Z as a subset of X, however as it occurs in the axiom $(a \to 0) \to 0 = a$ and the implication is meet-preserving in the second coordinate, it is more convenient to describe it by the collection of meet-irreducibles above it, *i.e.* by a subset of Y.

By Corollary 6.14 in Dunn, Gehrke, and Palmigiano (2005), the fusion in $\mathcal{G}(F)$ is associative iff F satisfies Φ_a:

$$\forall x_1, x_2, x_3 \quad \forall y$$
$$([\forall x_2'(\forall y'[R(x_2, x_3, y') \Rightarrow x_2' \leq y'] \Rightarrow R(x_1, x_2', y))]$$
$$\Leftrightarrow [\forall x_1'(\forall y''[(R(x_1, x_2, y'') \Rightarrow x_1' \leq y''] \Rightarrow R(x_1', x_3, y))])$$

Furthermore, by Corollary 6.17 in Dunn, Gehrke, and Palmigiano (2005), the fusion in $\mathcal{G}(F)$ is commutative iff F satisfies Φ_c:

$$\forall x_1, x_2 \; \forall y \; (R(x_1, x_2, y) \quad \Leftrightarrow \quad R(x_2, x_1, y))$$

For U to be the unit of the fusion in $\mathcal{G}(F)$ we have to ensure that $W \otimes U = W$ for all $W \in \mathcal{G}(F)$. As the fusion on $\mathcal{G}(F)$ is completely join-preserving, it suffices to ensure $x \otimes U = x$ for all $x \in X(= J^\infty(\mathcal{G}(F)))$. Note that,

$$
\begin{aligned}
x \otimes U \leq y \quad &\Leftrightarrow \quad \bigvee \{x \otimes x' \mid x' \leq U\} \leq y \\
&\Leftrightarrow \quad \forall x' \in U. x \otimes x' \leq y \\
&\Leftrightarrow \quad \forall x' \in U. R[x, x', \_]^l \subseteq \{y\}^l \\
&\Leftrightarrow \quad \forall x' \in U. y \in R[x, x', \_]^{lu} = R[x, x', \_] \quad \text{(as } R \text{ is compatible)} \\
&\Leftrightarrow \quad U \subseteq R[x, \_, y].
\end{aligned}
$$

Hence, U is the unit of the fusion in $\mathcal{G}(F)$ iff F satisfies Φ_u:

$$\forall x \, \forall y. \quad x \lessdot y \quad \Leftrightarrow \quad U \subseteq R[x, \_, y].$$

Now we have come to the last axiom: $(a \to 0) \to 0 = a$. First note that, by the adjunction property,

$$a \leq (a \to 0) \to 0 \quad \Leftrightarrow \quad (a \to 0) \otimes a \leq 0 \quad \Leftrightarrow \quad a \to 0 \leq a \to 0.$$

So in any case $a \leq (a \to 0) \to 0$. Furthermore, the mapping $a \mapsto (a \to 0) \to 0$ is completely join-preserving and therefore it again suffices to consider completely join-irreducible elements. Note that, for $x' \in \mathcal{J}^\infty(\mathcal{G}(\mathbf{F}))$,

$$
\begin{aligned}
x' \leq (x \to 0) \to 0 \quad &\Leftrightarrow \quad (x \to 0) \otimes x' \leq 0 \\
&\Leftrightarrow \quad x \to 0 \leq x' \to 0 \\
&\Leftrightarrow \quad \forall x''. x'' \leq x \to 0 \Rightarrow x'' \leq x' \to 0 \\
&\Leftrightarrow \quad \forall x''. x \otimes x'' \leq 0 \Rightarrow x' \otimes x'' \leq 0 \\
&\Leftrightarrow \quad \forall x''. Z \subseteq R[x, x'', \_] \Rightarrow Z \subseteq R[x', x'', \_]
\end{aligned}
$$

Hence, the equation $(a \to 0) \to 0 = a$ holds in $\mathcal{G}(F)$ iff F satisfies Φ_{dd}:

$$\forall x, x' \, (\forall x''. Z \subseteq R[x, x'', \_] \Rightarrow Z \subseteq R[x', x'', \_]) \Rightarrow x' \leq x.^5$$

Theorem 4.3. *The class of extended RS-frames* $F = (X, Y, \lessdot, R, U, Z)$ *satisfying* Φ_a, Φ_c, Φ_u *and* Φ_{dd} *gives complete semantics for MALL.*

---

[5]Note that the statement $x' \leq x$ uses the ordering of $\mathcal{G}(F)$. We may also write this in the language of the frame as: $\forall y. x \lessdot y \Rightarrow x' \lessdot y$.

Up to now we have computed the conditions on the relational frames correspond-ing to the axioms in a mechanical way, not worrying about getting the simplest pos-sible formulation. As our axioms could all be reduced to statements concerning join-irreducibles elements, these mechanical translations yield first-order statements on the dual. This illustrates the strength of using duality theory in the search for relational semantics: it allows a modular and uniform treatment of additional operations and ax-ioms. In the next section we will see that we may rewrite the conditions to get a cleaner representation and we will show that the semantics are closely related to phase spaces which are traditionally used as semantics for (multiplicative additive) linear logic.

5 Relational Frames and Phase Semantics

The traditional models used for MALL are so-called phase spaces. A *phase space* is a tuple $(M, \cdot, 1, \perp)$ where $(M, \cdot, 1)$ is a commutative monoid and $\perp \subseteq M$. One defines an operation on subsets A of M by

$$A^{\perp} = \{m \mid \forall n \in A. \, m \cdot n \in \perp\}. \tag{2.3}$$

A *fact* is a subset $F \subseteq M$ such that $(F^{\perp})^{\perp} = F$. MALL is interpreted in phase spaces by assigning facts to the basic propositions and interpreting the connectives as operations on facts (Girard 1987). As, for $A, B \in \wp(M)$, $B \subseteq A^{\perp} \Leftrightarrow A \subseteq B^{\perp}$, the mapping $(\_)^{\perp}$ yields a Galois connection on $\wp(M)$ and the Galois-closed sets are exactly the facts. The operations on facts corresponding to the connectives of MALL turn this collection of facts into a CL-algebra $\mathcal{F}ct(M)$. An inequality of MALL-formulas holds in a phase space M iff it holds in the corresponding CL-algebra $\mathcal{F}ct(M)$.

We will now see how phase semantics relate to the semantics derived in the previ-ous section.

Proposition 5.1. *Let* **L** *be a perfect CL-algebra. The subposets* $\mathcal{J}^{\infty}(\mathbf{L})$ *and* $\mathcal{M}^{\infty}(\mathbf{L})$ *of* **L** *are dually order-isomorphic.*

Proof. We will show that $(\_)^{\perp}$ restricts to a map $\mathcal{J}^{\infty}(\mathbf{L}) \rightarrow \mathcal{M}^{\infty}(\mathbf{L})$. The claim then follows from the fact that this map is its own inverse and is order-reversing. Let $x \in \mathcal{J}^{\infty}(\mathbf{L})$ and $A \subseteq \mathbf{L}$ such that $x^{\perp} = \bigwedge A$. Then

$$x = (x^{\perp})^{\perp} = (\bigwedge A)^{\perp} = \bigvee \{a^{\perp} \mid a \in A\}.$$

As $x \in \mathcal{J}^{\infty}(\mathbf{L})$, there exists $a \in A$ s.t. $x = a^{\perp}$, whence $x^{\perp} = (a^{\perp})^{\perp} = a$. □

By the previous proposition, for a CL-algebra **L**, its completely join-irreducibles and its completely meet-irreducibles are dually order-isomorphic and therefore the algebra may be described by a one-sorted frame based on the set $\mathcal{J}^{\infty}(\mathbf{L})$. Note that, for $x_1, x_2 \in \mathcal{J}^{\infty}(\mathbf{L})$,

$$x_1 \leq x_2^{\perp} \Leftrightarrow x_1 \leq x_2 \rightarrow 0 \Leftrightarrow x_1 \otimes x_2 \leq 0.$$

Hence, the order relation between $\mathcal{J}^{\infty}(\mathbf{L})$ and $\mathcal{M}^{\infty}(\mathbf{L})$ is completely determined by the fusion and the constant 0. Furthermore, in any CL-algebra, $1 = 0^{\perp}$, hence 1 is definable from 0 and the linear negation.

For a CL-algebra **L** we define a (one-sorted) frame $\mathcal{F}_1(\mathbf{L}) = (X, R_\downarrow, Z_\downarrow)$, by $X = \mathcal{J}^\infty(\mathbf{L})$, $Z_\downarrow = \{x \in X \mid x \le 0\}$ and, for $x_1, x_2, x_3 \in X$,

$$R_\downarrow(x_1, x_2, x_3) \iff x_3 \le x_1 \otimes x_2.$$

Conversely, for an RS-frame[6] $P = (X, R_\downarrow, Z_\downarrow)$ we define a Galois connection on $\wp(X)$ by, for $A \in \wp(X)$,

$$A^\perp = \{x \in X \mid \forall a \in A.\ R_\downarrow[x, a, \_] \subseteq Z_\downarrow\}. \tag{2.4}$$

We define a fusion on $\mathcal{G}_1(P)$, the Galois-closed subsets of P, by

$$
\begin{aligned}
x_1 \otimes x_2 &= \bigvee R[x_1, x_2, \_] && \text{for all } x_1, x_2 \in X,\\
w_1 \otimes w_2 &= \bigvee\{x_1 \otimes x_2 \mid x_1 \le w_1, x_2 \le w_2\} && \text{for all } w_1, w_2 \in \mathcal{G}_1(P).
\end{aligned}
$$

For a CL-algebra **L**, the structures $\mathcal{F}(\mathbf{L}) = (X, Y, \le, R, U, Z)$ and $\mathcal{F}_1(\mathbf{L}) = (X, R_\downarrow, Z_\downarrow)$ are directely interdefinable. For example, for $x_1, x_2, x_3 \in X$,

$$
\begin{aligned}
R_\downarrow(x_1, x_2, x_3) &\iff \forall y \in Y.\ R[x_1, x_2, y] \Rightarrow x_3 \le y\\
&\iff x_3 \in R[x_1, x_2, \_]^l.
\end{aligned}
$$

This allows us to translate the conditions Φ_a, Φ_c, Φ_u and Φ_{dd} to statements about one-sorted frames. E.g., Φ_{dd} becomes the statement Φ'_{dd}:

$$\forall x, x'\ (\forall x''.\ R_\downarrow[x, x'', \_] \subseteq Z_\downarrow \Rightarrow R_\downarrow[x', x'', \_] \subseteq Z_\downarrow) \Rightarrow x' \le x.$$

Translation of the other statements is left to the reader. For a one-sorted RS frame P, the algebra $\mathcal{G}_1(P)$, with constants 1 and 0 defined in the evident way, is a CL-algebra iff P satisfies Φ'_a, Φ'_c, Φ'_u and Φ'_{dd}.

Theorem 5.2. *One-sorted RS-frames $(X, R_\downarrow, Z_\downarrow)$, satisfying Φ'_a, Φ'_c, Φ'_u and Φ'_{dd} give complete semantics for MALL. We will call these structures CL-frames.*

For a CL-frame $P = (X, R_\downarrow, Z_\downarrow)$ we may define a phase space (only lacking a unit for the multiplication[7]) by $M_P = \wp(X)$, $\perp_P = \downarrow Z_\downarrow = \{A \in \wp(X) \mid A \subseteq Z_\downarrow\}$ and, for all $A, B \in \wp(X)$,

$$A \cdot_P B = \bigcup\{R[a, b, \_] \mid a \in A, b \in B\}.$$

As P satisfies Φ'_c, $\cdot_P$ is commutative.

Lemma 5.3. *For all $\mathcal{A} \in \wp(\wp(X))$, if $\mathcal{A}$ is a fact, i.e. $(\mathcal{A}^\perp)^\perp = \mathcal{A}$, then $\mathcal{A}$ is a principal downset in $\wp(\wp(X))$. Furthermore, for all $A \in \wp(X)$, A is Galois-closed in P iff $\downarrow A$ is a fact in M_P.*

Proof. We denote both the map (2.3) on $\wp(M_P)$ and the map (2.4) on $\wp(X)$ by $(\_)^\perp$, as the reader may derive the intended meaning from the context. Note that, for $\mathcal{A} \in \wp(M_P)$,

$$
\begin{aligned}
\mathcal{A}^\perp &= \{B \in M_P \mid \forall A \in \mathcal{A}.\ B \cdot_P A \in \perp_P\}\\
&= \{B \in M_P \mid \forall A \in \mathcal{A}.\ B \cdot_P A \subseteq Z_\downarrow\}\\
&= \{B \in M_P \mid B \cdot_P \bigcup \mathcal{A} \subseteq Z_\downarrow\}\\
&= \{B \in M_P \mid B \subseteq (\bigcup \mathcal{A})^\perp\}\\
&= \downarrow((\bigcup \mathcal{A})^\perp).
\end{aligned}
$$

---

[6]The notions 'reduced' and 'separating' are defined for one-sorted frames, as in Section 2 for two-sorted frames, in such a way that they ensure that X embeds in $\mathcal{G}_1(F)$ as its completely join-irreducibles.

[7]This is not a big issue as 1 is definable from the linear negation and 0.

From which the first claim follows immediately. The second claim easily follows from
$((\downarrow A)^{\perp})^{\perp} = \downarrow((A^{\perp})^{\perp})$. □

Theorem 5.4. *The CL-algebras $\mathcal{G}_1(P)$ and $\mathcal{F}ct(M_P)$ are isomorphic.*

Proof. It follows from the previous lemma that the mapping $A \mapsto \downarrow A$ is a bijection
between the two underlying sets. It is left to the reader to check that this map preserves
the CL-structure. □

Using the previous theorem, completeness of the semantics of phase spaces may
be derived from completeness of CL-frames. It is not always possible to construct,
given a phase space M, a CL-frame P_M s.t. $\mathcal{G}_1(P_M) \cong \mathcal{F}ct(M)$, as the complete lattice
$\mathcal{F}ct(M)$ may not be perfect.

The main advantage of working with CL-frames, instead of phase spaces, is that
they are in a duality with perfect CL-algebras which enables a modular and uniform
treatment of additional axioms and operations. Furthermore, the phase space describing a specific CL-algebra (e.g. the Lindenbaum algebra used in the completeness
proof) is in general much larger than the corresponding CL-frame. This size difference is also visible in the proof of Theorem 5.4: the underlying set of the phase space
associated to a CL-frame $(X, R_\downarrow, Z_\downarrow)$ is $\wp(X)$.

References

Allwein, G. and J. M. Dunn (1993). "A Kripke Semantics for Linear Logic". In: *Journal of Symbolic Logic* 58, pp. 514–545.

Almeida, A. (2009). "Canonical Extensions and Relational Representations of Lattices with Negation". In: *Studia Logica* 91, pp. 171–199.

Blackburn, P., M. de Rijke, and Y. Venema (2001). *Modal Logic*. Cambridge Tracts in Theoretical Computer Science 53. Cambridge University Press.

Coumans, D., M. Gehrke, and L. van Rooijen (2012). "Relational semantics for full linear logic". In: *Journal of Applied Logic (forthcoming)*. Special issue 'Logic, categories, semantics'.

Dunn, J. M., M. Gehrke, and A. Palmigiano (2005). "Canonical Extensions and Relational Semantics of Some Substructural Logics". In: *Journal of Symbolic Logic* 70, pp. 713–740.

Gehrke, M. (2006). "Generalized Kripke Frames". In: *Studia Logica* 84, pp. 241–275.

Gehrke, M. and J. Harding (2001). "Bounded Lattice Expansions". In: *Journal of Algebra* 238, pp. 345–371.

Gehrke, M. and B. Jónsson (1994). "Bounded Distributive Lattices with Operators". In: *Mathematica Japonica* 40, pp. 207–215.

Girard, J.-Y. (1987). "Linear Logic". In: *Theoretical Computer Science* 50, pp. 1–102.

Jónsson, B. and A. Tarski (1951). "Boolean Algebras with Operators, Part I". In: *American Journal of Mathematics* 73, pp. 891–939.

— (1952). "Boolean Algebras with Operators, Part II". In: *American Journal of Mathematics* 74, pp. 127–162.

Stone, M. (1936). "The Theory of Representations for Boolean Algebras". In: *Transactions of the American Mathematical Society* 40, pp. 37–111.

Troelstra, A. S. (1992). *Lectures on Linear Logic*. Stanford, CA: Center for the Study of Language and Information.

CHAPTER 3

Bivalent Logics

Vincent Degauquier

Abstract

This paper deals with bivalent predicate logics. In addition to classical logic, three bivalent logics, in which the principle of excluded middle and/or the principle of non-contradiction fail, can be distinguished. I provide a unified framework for studying the semantic and syntactic relationships between these four bivalent logics. More specifically, my purpose is to characterize the notion of logical consequence within each of these logics. To do this, I give a notion of validity and propose an associated sequent calculus.

1 Introduction

Logic is traditionally defined according to underlying principles. Among them, three seem particularly important. The principle of bivalence says that there are exactly two truth values, usually called True and False. The principle of excluded middle states that a sentence has at least one truth value. The principle of non-contradiction states that a sentence has at most one truth value. A logic that satisfies the conjunction of these three principles is called classical. By contrast, a logic is called non-classical if it does not obey at least one of them.

In relation to these principles, three bivalent logics differ from classical logic insofar as they ignore the principle of excluded middle and/or the principle of non-contradiction: consistent logic ignores the principle of excluded middle, complete logic ignores the principle of non-contradiction, and positive logic ignores both of these principles. In addition to classical logic, three bivalent (non-classical) logics can therefore be distinguished.

My purpose is to provide a unified framework for studying the semantic and syntactic relationships between these four bivalent logics. More specifically, my aim is to characterize the notion of logical consequence within each of these logics. To do this, I propose a new definition of the notions of model and sequent which makes these principles explicit.

The syntactic approach I have chosen is sequent calculus. For each of the logics mentioned above, I give a notion of validity and propose an associated sequent calculus. A sequent is called valid, glut-valid, gap-valid and classic-valid if it is semantically correct in positive logic, complete logic, consistent logic and classical logic, respectively. Similarly, a sequent is called derivable, glut-derivable, gap-derivable and classic-derivable if it is syntactically correct in positive logic, complete logic, consistent logic and classical logic, respectively.

2 Language

A first-order predicate language $\mathcal{L}$ is composed of a countable set of symbols consisting of a non-empty set of n-ary relation symbols, a set of n-ary function symbols, a countable set of variables and the usual logical symbols ($\neg$, $\wedge$, $\vee$, $\rightarrow$, $\forall$ and $\exists$). The nullary function symbols are called constants and the nullary relation symbols are called propositional symbols.

As for syntax, the notions of term and formula are defined in the usual way. Nevertheless, formulas equivalent up to bound variables are identified so that bound variables are supposed to be Bourbaki's squares. In this way, any occurrence of variable in a formula is free. The substitution operation is defined as follows. If A is a formula, α_1, ..., α_n are distinct variables and t_1, ..., t_n are terms, then $A[\alpha_1 := t_1, ..., \alpha_n := t_n]$ denotes the formula resulting from the simultaneous substitution of t_i for α_i in A, for all i ($1 \leq i \leq n$).

3 Semantics

A *positive model* $\mathcal{M}$ (simply called *model* hereafter) for a language $\mathcal{L}$ is composed of a structure for $\mathcal{L}$ and an interpretation of the proper symbols of $\mathcal{L}$ in this structure.

A *structure* for $\mathcal{L}$ consists of a universe, a set of relations on this universe and a set of functions defined on this universe and with values in this universe. For every $n \in \mathbb{N}$, if $\mathcal{L}$ has n-ary relation symbols, the structure must have at least one n-ary relation. For every $n \in \mathbb{N}$, if $\mathcal{L}$ has n-ary function symbols, the structure must have at least one n-ary function.

The universe $|\mathcal{M}|$ of a model $\mathcal{M}$ is a non-empty set. A n-ary relation R is an ordered pair of subsets of $|\mathcal{M}|^n$ such that $R = \langle R^+, R^- \rangle$. The first term of the ordered pair denotes the set of n-tuples of elements of the universe that verify the relation R

and the second term of the ordered pair denotes the set of n-tuples of elements of universe that falsify the relation.

An *interpretation* of $\mathcal{L}$ assigns to every constant an object in the universe, to every n-ary function symbol of $\mathcal{L}$ a n-ary function defined on the universe and to every n-ary relation symbol of $\mathcal{L}$ a n-ary relation, defined as an ordered pair of sets of n-tuples of elements of universe. The interpretation of a n-ary relation symbol R of $\mathcal{L}$ in the universe of the model $\mathcal{M}$ is denoted $R_{\mathcal{M}}$ and is equated to the ordered pair $\left\langle (R^n)^+_{\mathcal{M}}, (R^n)^-_{\mathcal{M}} \right\rangle$ of subsets of $|\mathcal{M}|^n$.

A *valuation* is an assignment of objects to variables. If v is a valuation, $\vec{\alpha}$ is a sequence of distinct variables $\alpha_1, ..., \alpha_n$ and $\vec{o}$ is a sequence of elements $o_1, ..., o_n$ in $|\mathcal{M}|$, then $v[\vec{\alpha} \mapsto \vec{o}]$ is the valuation that differs from v insofar as the variable α_i denotes the element o_i, for all i ($1 \leq i \leq n$). More specifically, the valuation $v[\vec{\alpha} \mapsto \vec{o}]$ is defined as follows :

- $v[\vec{\alpha} \mapsto \vec{o}](\beta) = o_i$, if β is the same variable as α_i ($1 \leq i \leq n$).

- $v[\vec{\alpha} \mapsto \vec{o}](\beta) = v(\beta)$, if β is distinct from every α_i ($1 \leq i \leq n$).

By combining a valuation v and an interpretation of constants and function symbols in $\mathcal{M}$, all terms of the language are given a value. The joint extension of interpretation and valuation is denoted $v_{\mathcal{M}}$. Moreover, if $\vec{t}$ is a sequence of terms $t_1, ..., t_n$, then $v_{\mathcal{M}}(\vec{t})$ denotes the sequence $v_{\mathcal{M}}(t_1), ..., v_{\mathcal{M}}(t_n)$. It is asked that :

- $v_{\mathcal{M}}(\beta) = v(\beta)$, for every variable β.

- $v_{\mathcal{M}}(F\vec{t}) = F_{\mathcal{M}}(v_{\mathcal{M}}(\vec{t}))$, for every n-ary function symbol F.

Truth and *falsity* of a formula A are defined in a model under a valuation. Given a model $\mathcal{M}$ and a valuation v, truth (denoted by $\mathcal{M} \models^+_v A$) and falsity (denoted by $\mathcal{M} \models^-_v A$) of formulas of the language are defined inductively :

$\mathcal{M} \models^+_v Rt_1...t_n$ if and only if $\langle v_{\mathcal{M}}(t_1), ..., v_{\mathcal{M}}(t_n) \rangle \in R^+_{\mathcal{M}}$

$\mathcal{M} \models^-_v Rt_1...t_n$ if and only if $\langle v_{\mathcal{M}}(t_1), ..., v_{\mathcal{M}}(t_n) \rangle \in R^-_{\mathcal{M}}$

$\mathcal{M} \models^+_v \neg A$ if and only if $\mathcal{M} \models^-_v A$

$\mathcal{M} \models^-_v \neg A$ if and only if $\mathcal{M} \models^+_v A$

$\mathcal{M} \models^+_v (A \wedge B)$ if and only if $\mathcal{M} \models^+_v A$ and $\mathcal{M} \models^+_v B$

$\mathcal{M} \models^-_v (A \wedge B)$ if and only if $\mathcal{M} \models^-_v A$ and/or $\mathcal{M} \models^-_v B$

$\mathcal{M} \models^+_v (A \vee B)$ if and only if $\mathcal{M} \models^+_v A$ and/or $\mathcal{M} \models^+_v B$

$\mathcal{M} \models^-_v (A \vee B)$ if and only if $\mathcal{M} \models^-_v A$ and $\mathcal{M} \models^-_v B$

$\mathcal{M} \models^+_v (A \rightarrow B)$ if and only if $\mathcal{M} \models^-_v A$ and/or $\mathcal{M} \models^+_v B$

$\mathcal{M} \models^-_v (A \rightarrow B)$ if and only if $\mathcal{M} \models^+_v A$ and $\mathcal{M} \models^-_v B$

$\mathcal{M} \vDash_v^+ \forall \alpha\, A$ if and only if $\mathcal{M} \vDash_{v[\alpha \mapsto o]}^+ A$, for all $o \in |\mathcal{M}|$

$\mathcal{M} \vDash_v^- \forall \alpha\, A$ if and only if $\mathcal{M} \vDash_{v[\alpha \mapsto o]}^- A$, for some $o \in |\mathcal{M}|$

$\mathcal{M} \vDash_v^+ \exists \alpha\, A$ if and only if $\mathcal{M} \vDash_{v[\alpha \mapsto o]}^+ A$, for some $o \in |\mathcal{M}|$

$\mathcal{M} \vDash_v^- \exists \alpha\, A$ if and only if $\mathcal{M} \vDash_{v[\alpha \mapsto o]}^- A$, for all $o \in |\mathcal{M}|$

Remark 3.1. *Although the notation might suggest it, the definitions of the connectives listed above are not equivalent to the usual definitions. By distinguishing truth from non-falsity and falsity from non-truth, the positive definitions of truth and falsity make the traditional definitions of logical connectives ambiguous. While it does not exist in classical logic, this ambiguity must be removed in a non-classical bivalent logic. It is therefore possible to propose several positive definitions corresponding to the classical definition of a connective but none of them fits perfectly with it. This translation problem is particularly acute in the cases of negation and implication.*

A model $\mathcal{M}$ is *consistent* if and only if $(R^n)_\mathcal{M}^+ \cap (R^n)_\mathcal{M}^- = \emptyset$, for every n-ary relation R on $|\mathcal{M}|$. A model $\mathcal{M}$ is *complete* if and only if $(R^n)_\mathcal{M}^+ \cup (R^n)_\mathcal{M}^- = |\mathcal{M}|^n$, for every n-ary relation R on $|\mathcal{M}|$. In this sense, a model is called *classical* if and only if it is both consistent and complete.

Proposition 3.2 (EXCLUDED MIDDLE). *Let $\mathcal{M}$ be a complete model.*

1. *if $\mathcal{M} \nvDash_v^+ A$, then $\mathcal{M} \vDash_v^- A$, for all formulas A.*

2. *if $\mathcal{M} \nvDash_v^- A$, then $\mathcal{M} \vDash_v^+ A$, for all formulas A.*

Proof. By induction on the complexity of A. □

Proposition 3.3 (NON-CONTRADICTION). *Let $\mathcal{M}$ be a consistent model.*

1. *if $\mathcal{M} \vDash_v^+ A$, then $\mathcal{M} \nvDash_v^- A$, for all formulas A.*

2. *if $\mathcal{M} \vDash_v^- A$, then $\mathcal{M} \nvDash_v^+ A$, for all formulas A.*

Proof. By induction on the complexity of A. □

Depending on whether a bivalent logic restricts the class of models to that of consistent, complete or classical models, this logic will be called consistent, complete or classical, respectively. In general, bivalent logic that takes into account the class of models without restriction is called positive.

4 Sequent Calculi

A *sequent* is a quadruple $\langle \Pi, \Gamma, \Delta, \Sigma \rangle$, where Π, Γ, Δ and Σ are finite sets of formulas of the language. The sequent $\langle \Pi, \Gamma, \Delta, \Sigma \rangle$ is denoted $\Pi; \Gamma \Vdash \Delta; \Sigma$. In addition, Γ, Γ' denotes $\Gamma \cup \Gamma'$ and A denotes $\{A\}$.

A sequent $\Pi; \Gamma \Vdash \Delta; \Sigma$ is *valid* if and only if for every model $\mathcal{M}$ and valuation v, $\mathcal{M} \nvDash_v^- A$, for all $A \in \Pi$, and $\mathcal{M} \vDash_v^+ A$, for all $A \in \Gamma$, implies $\mathcal{M} \vDash_v^+ A$, for some $A \in \Delta$, and/or $\mathcal{M} \nvDash_v^- A$, for some $A \in \Sigma$.

The definition of validity can be preserved for consistent and/or complete logics. Depending on whether the notion of valid sequent is restricted to consistent models or to complete models, a sequent is called *gap-valid* or *glut-valid*, respectively. If only the class of models which are both consistent and complete is taken into account, then a sequent is called *classic-valid*.

For each bivalent logic, a sequent calculus and a notion of derivability (corresponding to that of validity) are now set out. The rules of inference for these sequent calculi are as follows.

$$\frac{\Pi; \Gamma \Vdash \Delta; A, \Sigma}{\Pi; \Gamma, \neg A \Vdash \Delta; \Sigma} \neg_L^i \qquad\qquad \frac{\Pi, A; \Gamma \Vdash \Delta; \Sigma}{\Pi; \Gamma \Vdash \neg A, \Delta; \Sigma} \neg_R^i$$

$$\frac{\Pi; \Gamma \Vdash A, \Delta; \Sigma}{\Pi, \neg A; \Gamma \Vdash \Delta; \Sigma} \neg_L^e \qquad\qquad \frac{\Pi; \Gamma, A \Vdash \Delta; \Sigma}{\Pi; \Gamma \Vdash \Delta; \neg A, \Sigma} \neg_R^e$$

$$\frac{\Pi; \Gamma, A, B \Vdash \Delta; \Sigma}{\Pi; \Gamma, (A \wedge B) \Vdash \Delta; \Sigma} \wedge_L^i \qquad\qquad \frac{\Pi; \Gamma \Vdash A, \Delta; \Sigma \qquad \Pi; \Gamma \Vdash B, \Delta; \Sigma}{\Pi; \Gamma \Vdash (A \wedge B), \Delta; \Sigma} \wedge_R^i$$

$$\frac{\Pi, A, B; \Gamma \Vdash \Delta; \Sigma}{\Pi, (A \wedge B); \Gamma \Vdash \Delta; \Sigma} \wedge_L^e \qquad\qquad \frac{\Pi; \Gamma \Vdash \Delta; A, \Sigma \qquad \Pi; \Gamma \Vdash \Delta; B, \Sigma}{\Pi; \Gamma \Vdash \Delta; (A \wedge B), \Sigma} \wedge_R^e$$

$$\frac{\Pi; \Gamma, A \Vdash \Delta; \Sigma \qquad \Pi; \Gamma, B \Vdash \Delta; \Sigma}{\Pi; \Gamma, (A \vee B) \Vdash \Delta; \Sigma} \vee_L^i \qquad\qquad \frac{\Pi; \Gamma \Vdash A, B, \Delta; \Sigma}{\Pi; \Gamma \Vdash (A \vee B), \Delta; \Sigma} \vee_R^i$$

$$\frac{\Pi, A; \Gamma \Vdash \Delta; \Sigma \qquad \Pi, B; \Gamma \Vdash \Delta; \Sigma}{\Pi, (A \vee B); \Gamma \Vdash \Delta; \Sigma} \vee_L^e \qquad\qquad \frac{\Pi; \Gamma \Vdash \Delta; A, B, \Sigma}{\Pi; \Gamma \Vdash \Delta; (A \vee B), \Sigma} \vee_R^e$$

$$\frac{\Pi; \Gamma \Vdash \Delta; A, \Sigma \qquad \Pi; \Gamma, B \Vdash \Delta; \Sigma}{\Pi; \Gamma, (A \rightarrow B) \Vdash \Delta; \Sigma} \rightarrow_L^i \qquad\qquad \frac{\Pi, A; \Gamma \Vdash B, \Delta; \Sigma}{\Pi; \Gamma \Vdash (A \rightarrow B), \Delta; \Sigma} \rightarrow_R^i$$

$$\frac{\Pi; \Gamma \Vdash A, \Delta; \Sigma \qquad \Pi, B; \Gamma \Vdash \Delta; \Sigma}{\Pi, (A \rightarrow B); \Gamma \Vdash \Delta; \Sigma} \rightarrow_L^e \qquad\qquad \frac{\Pi; \Gamma, A \Vdash \Delta; B, \Sigma}{\Pi; \Gamma \Vdash \Delta; (A \rightarrow B), \Sigma} \rightarrow_R^e$$

$$\frac{\Pi; \forall \alpha A, \Gamma, A[\alpha := t] \Vdash \Delta; \Sigma}{\Pi; \Gamma, \forall \alpha A \Vdash \Delta; \Sigma} \forall_L^i \qquad\qquad \frac{\Pi; \Gamma \Vdash A[\alpha := \beta], \Delta; \Sigma}{\Pi; \Gamma \Vdash \forall \alpha A, \Delta; \Sigma} \forall_R^i$$

$$\frac{\forall \alpha A, \Pi, A[\alpha := t]; \Gamma \Vdash \Delta; \Sigma}{\Pi, \forall \alpha A; \Gamma \Vdash \Delta; \Sigma} \forall_L^e \qquad\qquad \frac{\Pi; \Gamma \Vdash \Delta; A[\alpha := \beta], \Sigma}{\Pi; \Gamma \Vdash \Delta; \forall \alpha A, \Sigma} \forall_R^e$$

$$\frac{\Pi; \Gamma, A[\alpha := \beta] \Vdash \Delta; \Sigma}{\Pi; \Gamma, \exists \alpha A \Vdash \Delta; \Sigma} \exists_L^i \qquad\qquad \frac{\Pi; \Gamma \Vdash A[\alpha := t], \Delta, \exists \alpha A; \Sigma}{\Pi; \Gamma \Vdash \exists \alpha A, \Delta; \Sigma} \exists_R^i$$

$$\frac{\Pi, A[\alpha := \beta]; \Gamma \Vdash \Delta; \Sigma}{\Pi, \exists \alpha A; \Gamma \Vdash \Delta; \Sigma} \exists_L^e \qquad\qquad \frac{\Pi; \Gamma \Vdash \Delta; A[\alpha := t], \Sigma, \exists \alpha A}{\Pi; \Gamma \Vdash \Delta; \exists \alpha A, \Sigma} \exists_R^e$$

The usual restrictions for $\forall_R^i$, $\forall_R^e$, $\exists_L^i$ and $\exists_L^e$ rules hold. The *eigenvariable* β must not appear in the conclusion of these rules.

The notion of *derivation* as well as those of *initial sequent* and *endsequent* are defined inductively as follows.

- A sequent $\Pi; \Gamma \Vdash \Delta; \Sigma$ is a derivation whose initial sequent and endsequent is $\Pi; \Gamma \Vdash \Delta; \Sigma$.

- If

$$\vdots$$
$$\Pi_1; \Gamma_1 \Vdash \Delta_1; \Sigma_1$$

is a derivation whose set of initial sequents is I_1 and whose endsequent is $\Pi_1; \Gamma_1 \Vdash \Delta_1; \Sigma_1$ and if

$$\frac{\Pi_1; \Gamma_1 \Vdash \Delta_1; \Sigma_1}{\Pi; \Gamma \Vdash \Delta; \Sigma} R$$

is an instance of the one-premiss rule R, then

$$\vdots$$
$$\frac{\Pi_1; \Gamma_1 \Vdash \Delta_1; \Sigma_1}{\Pi; \Gamma \Vdash \Delta; \Sigma} R$$

is a derivation whose set of initial sequents is I_1 and whose endsequent is $\Pi; \Gamma \Vdash \Delta; \Sigma$.

- If

$$\vdots \qquad\qquad \text{and} \qquad\qquad \vdots$$
$$\Pi_1; \Gamma_1 \Vdash \Delta_1; \Sigma_1 \qquad\qquad\qquad \Pi_2; \Gamma_2 \Vdash \Delta_2; \Sigma_2$$

are derivations whose sets of initial sequents are respectively I_1 and I_2, and whose endsequents are respectively $\Pi_1; \Gamma_1 \Vdash \Delta_1; \Sigma_1$ and $\Pi_2; \Gamma_2 \Vdash \Delta_2; \Sigma_2$ and if

$$\frac{\Pi_1; \Gamma_1 \Vdash \Delta_1; \Sigma_1 \qquad \Pi_2; \Gamma_2 \Vdash \Delta_2; \Sigma_2}{\Pi; \Gamma \Vdash \Delta; \Sigma} R$$

is an instance of the two-premiss rule R, then

$$\frac{\Pi_1; \Gamma_1 \Vdash \Delta_1; \Sigma_1 \qquad \Pi_2; \Gamma_2 \Vdash \Delta_2; \Sigma_2}{\Pi; \Gamma \Vdash \Delta; \Sigma} R$$

is a derivation whose set of initial sequents is $I_1 \cup I_2$ and whose endsequent is $\Pi; \Gamma \Vdash \Delta; \Sigma$.

A sequent is *derivable* if and only if there exists a derivation of which it is the endsequent and whose initial sequents are all axiomatic. A sequent $\Pi; \Gamma \Vdash \Delta; \Sigma$ is *axiomatic* if and only if $\Gamma \cap \Delta \neq \emptyset$ and/or $\Pi \cap \Sigma \neq \emptyset$.

The definition of derivable sequent can be preserved for consistent and/or complete logics by changing the definition of axiomatic sequent.

A sequent is *gap-derivable* if and only if there exists a derivation of which it is the endsequent and whose initial sequents are all gap-axiomatic. A sequent $\Pi; \Gamma \Vdash \Delta; \Sigma$ is *gap-axiomatic* if and only if it is axiomatic and/or $\Gamma \cap \Sigma \neq \emptyset$.

A sequent is *glut-derivable* if and only if there exists a derivation of which it is the endsequent and whose initial sequents are all glut-axiomatic. A sequent $\Pi; \Gamma \Vdash \Delta; \Sigma$ is *glut-axiomatic* if and only if it is axiomatic and/or $\Pi \cap \Delta \neq \emptyset$.

Finally, a sequent is *classic-derivable* if and only if there exists a derivation of which it is the endsequent and whose initial sequents are all classic-axiomatic. A sequent $\Pi; \Gamma \Vdash \Delta; \Sigma$ is *classic-axiomatic* if and only if it is gap-axiomatic and/or glut-axiomatic.

5 Some Results

Starting with the unified framework outlined above, it is now easy to prove the following propositions. While Propositions 5.1 and 5.2 show a hierarchy of bivalent logics theories, Propositions 5.3 and 5.5 show that the properties of derivability and classic-derivability are not reducible to those of glut-derivability and gap-derivability. Similarly, it is equally easy to prove the semantic results (involving the notions of validity, glut-validity, gap-validity and classic-validity) corresponding to these propositions.

Proposition 5.1. *If a sequent is derivable, then it is both glut-derivable and gap-derivable.*

Proposition 5.2. *If a sequent is glut-derivable and/or gap-derivable, then it is classic-derivable.*

Proposition 5.3. *Some sequents which are both glut-derivable and gap-derivable are not derivable.*

Example 5.4. *The sequent* ; $p \wedge \neg p \Vdash q \vee \neg q$; *is glut-derivable and gap-derivable but not derivable.*

$$
\cfrac{\cfrac{\cfrac{\cfrac{q\,;\,p \Vdash q\,;\,p}{;\,p \Vdash q,\neg q\,;\,p}\ \neg_R^i}{;\,p,\neg p \Vdash q,\neg q\,;}\ \neg_L^i}{;\,p,\neg p \Vdash q \vee \neg q\,;}\ \vee_R^i}{;\,p \wedge \neg p \Vdash q \vee \neg q\,;}\ \wedge_L^i
$$

Proposition 5.5. *Some classic-derivable sequents are neither glut-derivable nor gap-derivable.*

Example 5.6. *The sequent* $p \vee q$; $r \Vdash p$; $q \wedge r$ *is classic-derivable but neither glut-derivable nor gap-derivable.*

$$
\cfrac{\cfrac{p\,;\,r \Vdash p\,;\,q \qquad q\,;\,r \Vdash p\,;\,q}{p \vee q\,;\,r \Vdash p\,;\,q}\ \vee_L^e \qquad \cfrac{p\,;\,r \Vdash p\,;\,r \qquad q\,;\,r \Vdash p\,;\,r}{p \vee q\,;\,r \Vdash p\,;\,r}\ \vee_L^e}{p \vee q\,;\,r \Vdash p\,;\,q \wedge r}\ \wedge_R^e
$$

In set-theoretical terms, the conjunction of Propositions 5.1 and 5.3 asserts that the class of derivable sequents is strictly included in the intersection of the class of glut-derivable sequents and the class of gap-derivable sequents. As for Propositions 5.2 and 5.5, they assert that the union of the class of glut-derivable sequents and the class of gap-derivable sequents is strictly included in that of classic-derivable sequents.

Conclusion

The metatheoretical relationships between bivalent logics can be tackled either in terms of semantic generality or in terms of deductive power.

From the viewpoint of underlying principles, positive logic is more general than consistent and/or complete logics. These are defined from positive logic only by restricting the class of models. Understood positively, consistent and complete logics are nothing more than special cases of positive logic and classical logic is nothing more than a special case of both consistent logic and complete logic.

From the viewpoint of the correctness of sequents, it is well known that the class of correct sequents in non-classical logic is usually included in that of correct sequents in classical logic. Indeed, the class of glut-derivable and/or gap-derivable sequents is strictly included in the class of classic-derivable sequents. In addition, the class of derivable sequents is strictly included in that of sequents which are both glut-derivable and gap-derivable.

Thus, according to the viewpoint embraced, relationships between bivalent logics are understood in different ways. Nevertheless, in general, it can be said that a bivalent logic is deductively more powerful than another if and only if it is less general. Therefore, the positive interpretation of classical logic suggests a unified approach to bivalent logics that underlines the trade-off between a requirement of generality about truth and falsity and a requirement of deductive power.

References

Crabbé, M. (1992). "Soyons Positifs. La Complétude de la Théorie Naïve des Ensembles". In: *L'Anti-Fondation en Logique et en Théorie des Ensembles*. Ed. by R. Hinnion. Cahiers du Centre de Logique 7. Louvain-la-Neuve: Academia-Bruylant, pp. 51–68.

Gentzen, G. (1935). "Untersuchungen über das logische Schließen. I". In: *Mathematische Zeitschrift* 39, pp. 176–210.

Girard, J.-Y. (1976). "Three-Valued Logic and Cut-Elimination: the Actual Meaning of Takeuti's Conjecture". In: *Dissertationes Mathematicae (Rozprawy Matematyczne)* 136, pp. 1–49.

Muskens, R. (1999). "On Partial and Paraconsistent Logics". In: *Notre Dame Journal of Formal Logic* 40, pp. 352–374.

CHAPTER 4

Reducibility by Continuous Functions and Wadge Degrees

Kevin Fournier

Abstract

We consider the notion of reducibility by a continuous function, in particular for the Baire space. Reviewing the seminal results obtained mainly in the 70's, we describe the Wadge hierarchy for a topological class Γ that has good closure and determinacy properties, proving that it is quasi-well-ordering. Finally, we state some results for the real line.

1 Introduction

Descriptive set theory is in essence the study of the complexity of subsets of the Baire space ω^ω, the 'logician's reals'. This work is devoted to a natural measure of the relative complexity of subsets of the Baire space, namely *reducibility by a continuous function*. Given two subsets A and B of the Baire space, A is said to be reducible to B, and we write $A \leq_W B$, if and only if A is the preimage of B for some continuous function f from the Baire space to itself. If we understand the complexity of A to mean the difficulty of determining membership in A, we observe that if A is reducible to B then A is, in a certain sense, no more complicated than B. Another motivation for studying reducibility by a continuous function can be found in the equivalence relation induced by $\leq_W$. These equivalence classes, called the *Wadge degrees*, are indeed the smallest classes of subsets which are closed under continuous preimages. In this sense, Wadge degrees can be considered as a refinement of the well-known Borel and projective classes. The relation $\leq$ on Wadge degrees is, merely by definition, a partial preordering called the *Wadge order*. If we restrict ourselves to a class Γ of subsets of the Baire space with appropriate closure and determinacy properties (e.g. Borel sets or projective sets under projective determinacy), it is in fact a well-quasi-ordering on Wadge degrees. This follows from two important results: Wadge's Lemma (Wadge 1983) and the Martin-Monk Theorem (Van Wesep 1978). Both are proved using a very powerful correspondence between reducibility by a continuous function and a certain infinite two-player game called the *Wadge Game*. In contrast, for topological spaces that are not zero-dimensional, the structure of the Wadge order may be more chaotical (Schlicht 2010). The situation is completely different from the case of the Baire space and it is not possible to get the same kind of game characterization. For example, Wadge's Lemma fails and the Wadge order for the real line is ill-founded.

2 Wadge Games

We recall that a tree T on a nonempty set Λ is a subset of $\Lambda^{<\omega}$ that is closed under subsequences. It is pruned if every element in the tree admits a strict extension that is also in the tree. When T is some tree, we denote by $[T]$ the set of all its infinite branches:

$$[T] = \{x \in \Lambda^\omega \mid \forall n \in \omega \, (x|n \in T)\}.$$

The topology on Λ^ω is the product topology of the discrete topology on Λ: $A \subseteq \Lambda^\omega$ is open if and only if there exists some $U \subseteq \Lambda^{<\omega}$ such that

$$A = \{u\hat{\ }a \in \Lambda^\omega \mid u \in U\},$$

where the symbol $\hat{\ }$ denotes the concatenation. In this paper, we will only consider the case of the Baire space, where $\Lambda = \omega$. Nevertheless, the main results are general. To begin, we state a fundamental lemma[1] on continuous maps of the Baire space.

Lemma 2.1. *Let $\varphi \colon \omega^{<\omega} \longrightarrow \omega^{<\omega}$ be* monotone, *i.e. such that for any $s, t \in \omega^{<\omega}$:*

$$s \subseteq t \longrightarrow \varphi(s) \subseteq \varphi(t),$$

and proper, *i.e. such that for every $x \in \omega^\omega$*

$$\lim_{n \to \infty} \varphi(x|n) \in \omega^\omega.$$

Then the application

$$\varphi^* \colon \omega^\omega \longrightarrow \omega^\omega$$
$$x \longmapsto \bigcup_{n \in \omega} \varphi(x|n).$$

is continuous. Conversely, for every continuous function f from the Baire space to itself, there exists a monotone and proper map $\varphi \colon \omega^{<\omega} \longrightarrow \omega^{<\omega}$ such that $f = \varphi^*$.

This lemma shows that the continuous functions from the Baire space to itself are exactly given by the monotone and proper maps on finite sequences.

---

[1] See for example Kechris (1994, p. 8).

Definition 2.2. *Let $A, B \subseteq \omega^\omega$. A continuous reduction of A to B is a continuous function $f\colon \omega^\omega \longrightarrow \omega^\omega$ such that $A = f^{-1}(B)$. It is denoted $A \leq_W B$, and A is said to be Wadge reducible to B. If $A \leq_W B \leq_W A$, then A is said to be Wadge-equivalent to B. If $A \leq_W B$, but A is not Wadge-equivalent to B, we write $A <_W B$.*

Intuitively, $A \leq_W B$ means that A is less complicated than B. Indeed, given $x \in \omega^\omega$ and a continuous reduction f of A to B, to determine if x is in A we only need to compute the image $f(x)$ of x by f, and determine whether it is in B or not. The problem of determining membership in A can therefore be reduced to that of determining membership in B, provided that the computation of $f(x)$ introduces no extra complexity.

Lemma 2.3. *The relation $\leq_W$ on subsets of ω^ω is a preorder.*

Proof. Let $A \subseteq \omega^\omega$. The identity function on ω^ω is a continuous function such that the inverse image of A is itself, so $A \leq_W A$, and that proves the reflexivity of $\leq_W$.

Let $A, B, C \subseteq \omega^\omega$ such that $A \leq_W B$ and $B \leq_W C$. Then there are continuous functions f and g from ω^ω to itself such that $A = f^{-1}(B)$ et $B = g^{-1}(C)$. We have then $A = f^{-1}(g^{-1}(C)) = (g \circ f)^{-1}(C)$. But $g \circ f$ is continuous, so $A \leq_W C$, which proves the transitivity of $\leq_W$.   □

Let $A, B \subseteq \omega^\omega$. We consider the *Wadge game* $W(A, B)$:

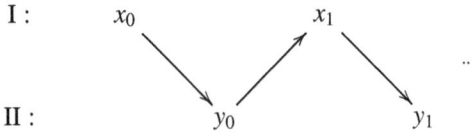

I : x_0 x_1

                                                             ...

II : y_0 y_1

In this infinite game, players play alternatively elements of ω, building two infinite sequences, $x = (x_0, x_1, \ldots) \in \omega^\omega$ for player I and $y = (y_0, y_1, \ldots) \in \omega^\omega$ for player II. Player II is allowed to skip, even ω times, provided she also plays ω times. She wins if and only if $(x \in A \Leftrightarrow y \in B)$, otherwise, I wins. These rules are designed so that II has a winning strategy in the Wadge game $W(A, B)$ if and only if $A \leq_W B$.

Proposition 2.4. *Let $A, B \subseteq \omega^\omega$, then II has a winning strategy in the Wadge game $W(A, B)$ if and only if $A \leq_W B$.*

Proof. A strategy for II in the Wadge game $W(A, B)$ is a map

$$\sigma\colon \omega^{<\omega \setminus \{0\}} \longrightarrow \omega^{<\omega}$$

such that:

1. for all $a, b \in \omega^{<\omega \setminus \{0\}}$ with $a \subseteq b$, $\sigma(a) \subseteq \sigma(b)$;

2. for all $x \in \omega^\omega$, $\lim_{n \to \infty} \sigma(x|n) \in \omega^\omega$.

During the game, the strategy tells II what she has to play, given the moves of her opponent. It may ask her to play a finite sequence of length greater than one, but in this case she just plays the first element of the sequence. As the game is infinite and the strategy consistent (because of the first condition), she will be able to follow the strategy until the end. The second condition insures that player II will indeed play an infinite sequence during the game.

A continuous function $\sigma^*: \omega^\omega \longrightarrow \omega^\omega$ arises thus from a strategy for II. If the strategy is winning, we have moreover that, for all $x \in \omega^\omega$, $x \in A \Leftrightarrow \sigma^*(x) \in B$, i.e. $(\sigma^*)^{-1}(B) = A$, so that $A \leq_W B$.

Conversely, if $A \leq_W B$, there exists a continuous function $\varphi: \omega^\omega \longrightarrow \omega^\omega$ such that $\varphi^{-1}(B) = A$. But this continuous function arises from a map on the finite sequences that can be seen as a strategy for II. Since $\varphi^{-1}(B) = A$, it is winning. ☐

This game characterization leads us to a very important result obtained by Wadge (1983), and often called *Wadge's Lemma*. Let Γ be a topological class of subsets of the Baire space, and denote by $AD_W(\Gamma)$ the fact that for every subsets A and B of the Baire space in Γ, the game $W(A, B)$ is determined, that is either player I or II has a winning strategy in this game.

Theorem 2.5 (Wadge's Lemma). *Let Γ be a topological class of subsets of the Baire space such that $AD_W(\Gamma)$ holds. Then for every $A, B \in \Gamma$, we have*

$$A \leq_W B \qquad or \qquad B \leq_W A^c.$$

Proof. If II has a winning strategy in $W(A, B)$, we have already seen that $A \leq_W B$. If it is I who has a winning strategy σ, it means that whatever plays II, the sequence played by I (following her winning strategy) will be in A^c if and only if the sequence played by II is in B. From the winning strategy σ for I in the game $W(A, B)$, we can then construct a winning strategy σ' for II in the game $W(B, A^c)$:

$$\sigma'(x_0) = \sigma(\emptyset);$$
$$\sigma'((x_0, x_1, \ldots, x_i, x_{i+1})) = \sigma((x_0, x_1, \ldots, x_i)).$$

Hence, $B \leq_W A^c$. ☐

The determinacy of Wadge games can be seen as a consequence of the determinacy of Gale-Stewart games for topological classes with appropriate closure properties. For instance, if Γ is the class of Borel subsets of the Baire space, $AD_W(\Gamma)$ holds in ZF. Under projective determinacy, $AD_W(\Gamma)$ holds for Γ the projective subsets of the Baire space; and under full determinacy we have $AD_W(\mathscr{P}(\omega^\omega))$.

3 Wadge Hierarchy

The relation $\equiv_W$ is an equivalence relation, and its classes denoted $[A]_W$ are called *Wadge degrees*. The relation $\leq_W$ induces a relation $\leq$ on Wadge degrees:

$$[A]_W \leq [B]_W \Longleftrightarrow A \leq_W B.$$

The relation $\leq$ is a partial order. Moreover, by definition, if $[A]_W \leq [B]_W$ and $[B]_W \leq [A]_W$, we have $A \leq_W B$ and $B \leq_W A$. So $A \equiv_W B$, and then $[A]_W = [B]_W$. Thus the relation $\leq$ is also antisymmetric.

Let Γ be a topological class of subsets of the Baire space such that $AD_W(\Gamma)$ holds. Wadge's Lemma tells us that for the partial order $\leq$, any antichain has a size at most 2: a given Wadge degree $[A]_W$ is comparable to any other Wadge degree, except $[A^c]_W$ if they are not equal. Indeed let $[B]_W$ be another Wadge degree; by the Wadge Lemma, we have:

$$A \leq_W B \qquad \text{or} \qquad B^c \leq_W A,$$

but also

$$B \leq_W A \qquad \text{or} \qquad A^c \leq_W B.$$

The only case when A and B are incomparable would be

$$A^c \leq_W B \qquad \text{and} \qquad B^c \leq_W A.$$

But $B^c \leq_W A$ is equivalent to $B \leq_W A^c$; we would have then $A^c \leq_W B$ and $B \leq_W A^c$, and thus $B \equiv_W A^c$, i.e. $[B]_W = [A^c]_W$.

Theorem 3.1 (Martin, Monk). *Let Γ be a topological class of subsets of the Baire space such that $AD_W(\Gamma)$ holds and every element in Γ has the Baire property (BP). Then the relation $\leq_W$ is well-founded on Γ.*

Proof. If an infinite descending chain existed in Γ, there would be $(A_n)_{n\in\omega} \subseteq \Gamma$ such that $\cdots <_W A_2 <_W A_1 <_W A_0$. Since for all $n \in \omega$, $A_n \nleq_W A_{n+1}$, and $A_n \nleq_W A^c_{n+1}$, the player I would have, for all n, two winning strategies: σ^0_n in $W(A_n, A_{n+1})$ and σ^1_n in $W(A_n, A^c_{n+1})$. Fixing $x = (x_0, x_1, \ldots) \in 2^\omega$, we could then construct the following diagram:

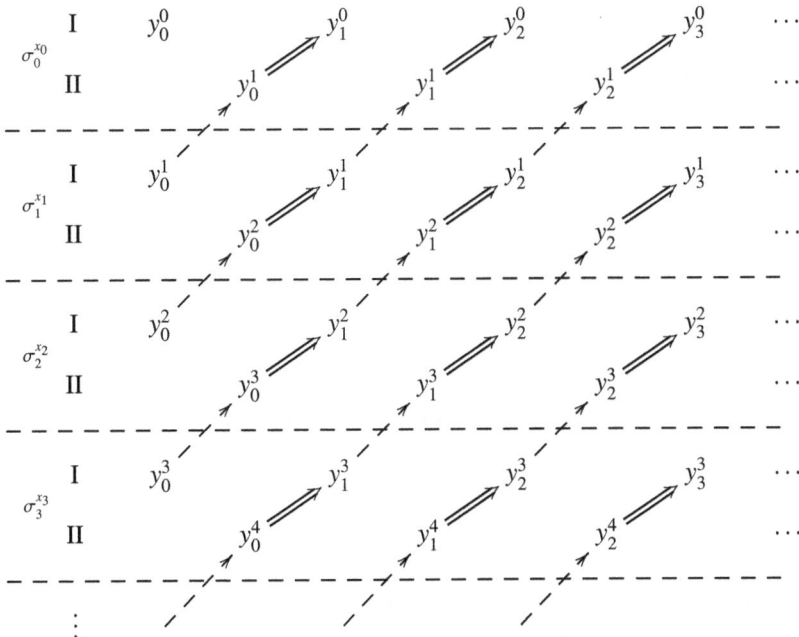

where the ($\Rightarrow$) arrows denote the elements played by I according to her strategy, and the ($\rightarrow$) arrows elements copied from one game to another. I plays y^n_0 in the nth game following $\sigma^{x_n}_n$, this fills the first column. Then II copies as shown to play y^{n+1}_0 in the nth game, this fills the second column. I responds by following $\sigma^{x_n}_n$ in the nth game, and so on.

Let $y_n(x) = (y_k^n)_{k\in\omega}$. Then, since the strategies $\sigma_n^{x_n}$ are winning for I, we have:

$$y_n(x) \notin A_n \Leftrightarrow y_{n+1}(x) \in A_{n+1}^{x_n},$$

where $A_n^0 = A_n$ and $A_n^1 = A_n^c$. Let

$$X = \{x \in 2^\omega \mid y_0(x) \in A_0\}.$$

Since $f: x \mapsto y_n(x)$ is continuous, $X = f^{-1}(A_0)$ is in Γ, and hence has the BP. Notice now that X is what we called a *flipset*: let $x, x' \in 2^\omega$, if x and x' are such that, for some $i \in \omega$, $x_k = x_k'$ for all $k \neq i$ and $x_i \neq x_i'$, then $x \in X \Leftrightarrow x' \notin X$. Since X has the BP, there is $n \in \omega$ and $s \in 2^n$ so that X is either meager or comeager in $s\hat{}2^\omega$. Let $\varphi: s\hat{}2^\omega \longrightarrow s\hat{}2^\omega$ be the homeomorphism given by $\varphi(x) = (x_0, x_1, \ldots, x_{n-1}, 1 - x_n, x_{n+1}, \ldots)$. Then $x \in X \Leftrightarrow \varphi(x) \notin X$, so that $\varphi(X \cap (s\hat{}2^\omega)) = (s\hat{}2^\omega\backslash X) \cap s\hat{}2^\omega$, which is a contradiction. □

This last theorem shows that the relation $\leq_W$ is well-founded on Γ, and hence proves that we really have a hierarchy: the *Wadge hierarchy*. The hypotheses of the theorem are for example fullfilled in ZF if Γ is the class of Borel subsets of the Baire space, or if Γ is the projective subsets of the Baire space under projective determinacy.

From now on, we only consider subsets of the Baire space in a topological class Γ with appropriate closure properties, such that $AD_W(\Gamma)$ holds and every element in Γ has the BP.

Definition 3.2. *A subset of the Baire space is* self-dual *if it is Wadge equivalent to its complement. A Wadge degree is* self-dual *if its elements are self-duals.*

Lemma 3.3. *Immediatly above a non self-dual degree $[A]_W$, there is a self-dual degree.*

Proof. Consider the set

$$B = \{0\}\hat{}A \cup \{1\}\hat{}A^c.$$

Intuitively, in terms of Wadge games, it means that the player in charge of the set B chooses at her first turn to play with A or with A^c. Notice that the set B is self-dual. Since A is non self-dual, II doesn't have a winning strategy in $W(A^c, A)$; hence, by determinacy, I has a winning strategy σ in this game. Thus $A <_W B$, because I has a winning strategy when she is in charge of B against A: she chooses A^c and then follows σ. To prove that its degree is immediately above $[A]_W$, let C be such that $C <_W B$. Then I has a winning strategy in the Wadge game when she is in charge of B against C. But as she is playing first, she must choose between A and A^c before any move of player II. If her winning strategy is to play with A, then II has a winning strategy in $W(C, A)$, and $C \leq_W A$. If her winning strategy is to play with A^c, then II has a winning strategy in $W(C, A^c)$, and $C \leq_W A^c$. □

Before going any further, we prove a theorem that will provide us a very useful characterization of non self-dual sets.

Theorem 3.4. *Let $A \subseteq \omega^\omega$ in Γ and for any $u \in \omega^{<\omega}$, define*

$$A[u] = \{\alpha \in \omega^\omega \mid u\hat{}\alpha \in A\}.$$

Then A is non-self-dual if and only if there exists $\alpha \in \omega^\omega$ such that for any $n \in \omega$, $A[\alpha \upharpoonright n] \equiv_W A$.

Proof. Suppose first that A is non-self-dual. For all $\alpha \in \omega^\omega$, II has a trivial winning strategy in $W(A[\alpha \restriction n], A)$ which is the following: II skips until I has played her first n moves: if I has played $\alpha \restriction n$, II skips one more time and then plays the same moves as I. If I does not have played $\alpha \restriction n$, then II plays moves to construct a sequence in A^c. Thus, $A[\alpha \restriction n] \leq_W A$. To show that $A \leq_W A[\alpha \restriction n]$, we first notice that, since A is non-self-dual, I has a winning strategy σ in the game $W(A^c, A)$. Consider the game when II begins to skip n times. Then σ provides I a sequence of length $n+1$ such that, whatever plays II next, I can still win. We construct that way an infinite sequence α that satisfies $A \leq_W A[\alpha \uparrow n]$ for all $n \in \omega$.

The proof of the converse is very close to the proof of Theorem 3.1, using a sequence of games toward a contradiction with the fact that sets in Γ have the BP. □

Lemma 3.5. *Immediatly above a self-dual degree* $[A]_W$, *there is a non self-dual degree.*

Proof. We define:

$$B = \bigcup_{m,n\in\omega} 0^m\!{}^\frown\{n+1\}^\frown A \qquad \text{and} \qquad C = \{0^\omega\} \cup B.$$

These sets are non self-dual, because the sequence 0^ω verifies $B[0^\omega \restriction n] \equiv_W B$ and $C[0^\omega \restriction n] \equiv_W C$ for any integer n. Moreover, $A \leq_W B$ and $A \leq_W C$. We show now that $B \equiv_W C^c$. First, we prove that $B \leq_W C^c$. Let σ be a winning strategy for II in $W(A, A^c)$, then II has the following winning strategy in $W(B, C^c)$: until I plays 0, II plays also 0, if I stops playing 0, II plays an integer and then follows σ. Similarly, we can prove that $C \leq_W B^c$, and thus $B \equiv_W C^c$.

To prove that $[B]_W$ and $[C]_W$ are immediatly above $[A]_W$, consider first D such that $D <_W B$. The player I then has a winning strategy σ in $W(B, D)$, and we construct from it a strategy for II in $W(D, A^c)$: II follows exactly σ, only replacing the $0^m\!{}^\frown\{n+1\}$ by skips. Thus $D \leq_W A$, and the same result holds if $D <_W C$. □

Since $[\emptyset]_W$ and $[\omega^\omega]_W$ are trivially non-self-dual and of minimal degree, we can now have a picture of the finite levels of the hierarchy. To know what happens at limit stages, we prove a consequence of Theorem 3.4 about self-dual degrees.

Proposition 3.6. *Let* $A \subseteq \omega^\omega$ *in* Γ *be self-dual, then there exists* $(A_i)_{i\in I}$, $I \subseteq \omega^{<\omega^*}$, *each* $A_i \subseteq \omega^\omega$ *non self-dual and* $A_i <_W A$ *such that:*

$$A \equiv_W \bigcup_{i\in I} i^\frown A_i.$$

Proof. By induction on the Wadge degrees, assume that the property holds for all self-dual B such that $B <_W A$. Set:

$$I_0 = \{\alpha \restriction n_\alpha \mid \alpha \in \omega^\omega \text{ and } n_\alpha \text{ is the least integer such that } A[\alpha \restriction n_\alpha] <_W A\}.$$

For $u \in I_0$, $A[u] <_W A$ and $A = \bigcup_{u\in I_0} u^\frown A[u]$; by induction hypothesis, if $A[u]$ is self-dual, $A[u] \equiv_W \bigcup_{i\in I_u} i^\frown A[u]_i$, with $A[u]_i \subseteq \omega^\omega$ non self-dual and $A[u]_i <_W A[u]$. Thus, setting $I_1 = \{u \in I_0 \mid A[u] \text{ is self-dual }\}$, and $I_2 = I_0 \backslash I_1$, we have

$$A \equiv_W \bigcup_{v\in I_2} v^\frown A[v] \cup \bigcup_{u\in I_1} u^\frown\left(\bigcup_{i\in I_u} i^\frown A[u]_i\right) = \bigcup_{v\in I_2} v^\frown A[v] \cup \bigcup_{u^\frown i\in I} u^\frown i^\frown A[u]_i,$$

where $I = \{u\hat{\ }i \mid u \in I_1, i \in I_u\}$.   □

The last proposition proves that, there is no self-dual sets at limit stages of cofinality greater or equal to ω_1.[2]

Proposition 3.7. *In the Wadge hierarchy, we find self-dual sets at limit stages of cofinality ω.*

Proof. Let $(A_n)_{n \in \omega} \subseteq \Gamma$ be such that for all $i \in \omega$, $A_i <_W A_i + 1$. We consider then the following set:

$$B = \bigcup_{i \in \omega} \hat{\imath} A_i.$$

By definition, we have for all $i \in \omega$, $A_i <_W B$. Moreover, let $C <_W B$, it means that I has a winning strategy in the game $W(B, C)$. In her first move, I chooses a set, say A_i, and then, whatever plays II, I wins the game. This winning strategy gives II a winning strategy in the game $W(C, A_i^c)$, so that $C \leq_W A_i^c$ and then $C <_W A_{i+1}$. Thus, $[B]_W$ is the least degree above all the $[A_i]_W$. Notice that:

$$B^c = \bigcup_{i \in \omega} \hat{\imath} A_i^c.$$

B is self-dual because II has a winning strategy in $W(B^c, B)$: if I plays i first, II answers by $i + 1$ and then follows the winning strategy given by $A_i^c <_W A_{i+1}$.   □

We can now have a global picture of the Wadge hierarchy on Γ:

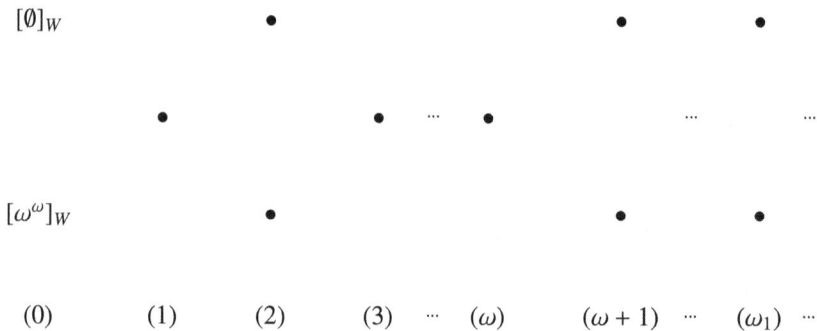

$[\emptyset]_W$ •   •   •

•   • ⋯ •   ⋯   ⋯

$[\omega^\omega]_W$ •   •   •

(0)   (1)   (2)   (3)   ⋯ (ω) $(\omega + 1)$ ⋯ (ω_1) ⋯

4 The Case of the Real Line

The structure of the Wadge order on topological spaces that are not zero-dimensional may be more chaotical. In this section, we will study the case of the real line, which is eloquent. As a game characterization is no longer possible, some results that were very easy to derive for the Baire space using the Wadge games are now very difficult to prove, nay impossible. For instance, Wadge's Lemma fails and the Wadge order is ill-founded.

Let $\mathbb{N} = \omega \setminus \{0\}$. For any $a, b \subseteq \mathscr{P}(\mathbb{N})$, we say that $a \subseteq_{fin} b$ if and only if $a \setminus b$ is finite. We can prove the following theorem.[3]

---

[2]This may change if the space considered is not the Baire space but Λ^ω, depending on the cardinality of Λ.

[3]Ikegami (2010, p. 114).

Theorem 4.1. *There is an embedding*

$$i\colon (\mathscr{P}(\mathbb{N}), \subseteq_{fin}) \longrightarrow (\mathscr{P}(\mathbb{R}), \leq_W)$$

such that the range of i consists of subsets of real numbers which are the difference of open sets.

As it is easy to construct an infinite descending sequence of subsets of $\mathbb{N}$ with respect to $\subseteq_{fin}$, we can, via the embedding i construct an infinite descending sequence of subsets of the real line with respect to the Wadge order. Hence, the Wadge order on the real line is ill-founded.

Moreover, using the embedding i, we can prove[4] that Wadge's Lemma fails for the real line.

Corollary 4.2. *There are two subsets of the real line A and B which are the difference of two open sets such that neither $A \leq_W B$ nor $B \leq_W A^c$ hold.*

References

Ikegami, D. (2010). "Games in Set Theory and Logic". PhD thesis. Institute for Logic, Language and Computation (ILLC), Universiteit van Amsterdam.

Kechris, A. S. (1994). *Classical Descriptive Set Theory*. New York: Springer Verlag.

Schlicht, P. (2010). "The Wadge Order for Polish spaces". Preprint.

Van Wesep, R. (1978). "Wadge Degrees and Descriptive Set Theory". In: *Cabal Seminar 76–77*. Ed. by A. S. Kechris and Y. N. Moschovakis. Lecture Notes in Mathematics 689. New York: Springer, pp. 151–170.

Wadge, W. (1983). "Reducibility and Determinateness on the Baire Space". PhD thesis. University of California, Berkeley.

---

[4] Ikegami (2010, p. 116).

CHAPTER 5

Abduction of Multiple Explanatory Hypotheses

Tjerk Gauderis

Abstract

In abduction—the process of finding explanatory hypotheses for puzzling phenomena—one is often confronted with multiple explanatory hypotheses. In science one generally wants to test further the different hypotheses one by one. But, if we try to model this in a logic and make it possible to derive the different hypotheses apart from each other, we generally can derive their conjunction too. An elegant solution within the framework of adaptive logics is provided in Gauderis (2011). But this approach is not restricted to science. While it is true that a lot of cases in everyday reasoning require a more practical approach—in which one acts on the knowledge that all the different hypotheses might be the case—there is also a considerable amount of situations in which the more theoretical approach of the scientist is needed. In this paper we try to illustrate this by using this logic to model reasoning within detective literature.

1 The Problem of Multiple Explanatory Hypotheses

Abduction and Detective literature. Charles Peirce thought that there were three characteristic ways of reasoning in science. In addition to the better-known ways of deduction and induction, there was a third rational way in which scientists can reason: abduction or "the process of forming an explanatory hypothesis" (Peirce 1958–60, CP 5.171). The logical schema of forming such an explanatory hypothesis is for Peirce (1958–60, CP 5.189) the following:

The surprising fact, C is observed;
But if A were true, C would be a matter of course,
Hence, there is reason to suspect that A is true.

Obviously, this is a form of defeasible or fallible reasoning, of which Peirce (1958–60, CP 2.777) himself was perfectly conscious: "The hypothesis which it problematically concludes is frequently utterly wrong itself, and even the method needs not ever lead to the truth." When we translate his schema to predicate logic, we get the following schema:

$$(\forall\alpha)(A(\alpha) \supset B(\alpha)), B(\beta)/A(\beta).$$

This schema is better known as the logical fallacy *Affirming the Consequent*, but this is only a fallacy if we stick within deductive logics. In cases in which we are not able to obtain any deductive results that can explain our observations, Peirce (1958–60, CP 2.777) makes a point in stating that "...its method is the only way in which there can be any hope of attaining a rational explanation".

Quite often, people have acknowledged that this is essentially the same kind of reasoning as the reasoning employed in crime investigation or detective literature. As the different articles in the book 'The Sign of Three'[1] point out, solving a murder case by tracing back the clues is essentially an abductive operation. So, while Holmes was maybe wrong when he said to Watson that it was elementary deduction, he surely was not wrong in thinking that his reasoning was logical, it was only according to the laws of a logic for abduction.

The Problem of Multiple Explanatory Hypotheses. Still, in solving a murder case, our detective is quite often confronted with two or more suspects. When we try to model this with a formal logic, this can lead to a problem. Consider the following example. Suppose we are confronted with the puzzling fact or clue Pa while our background knowledge contains two possible causes: $(\forall x)(Qx \supset Px)$ and $(\forall x)(Rx \supset Px)$. There are actually now two roads that can be taken. We could construct a logic in which we can only derive the disjunction $(Qa \vee Ra)$ and not the individual hypotheses Qa and Ra. This road, called *practical* abduction,[2] is suitable to model situations in which one has to *act* on the basis of the conclusions. For instance, in medical diagnoses, a physician who finds out that two possible diseases can be the cause for the examined symptoms, needs to take appropriate steps based on the fact that both diseases might be the cause.

But our detective has a more theoretical perspective and is interested in finding out which of the hypotheses is the actual cause. Therefore, it is important that he can *abduce* the individual hypotheses Qa and Ra in order to examine them further one by one. This is because, on the one hand, one has to be able to derive Qa and Ra separately, but on the other hand, one has to prevent the derivation of their conjunction $(Qa \wedge Ra)$. Not only does it seem counterintuitive to take the conjunction of

---

[1] See Sebeok and Eco (1988). This book investigates the relation between the writings of Charles Peirce on the one hand and the writings of Edgar Allen Poe (Auguste Dupin) and Arthur Conan Doyle (Sherlock Holmes) on the other hand.

[2] According to the definition suggested in Meheus and Batens (2006, pp. 224–225) and used in Lycke (2009).

two possible hypotheses as an explanation. Also, if the two hypotheses are actually incompatible – a victim cannot be murdered twice – it would lead to explosion. So it is clear that for this application, we need the second road.

Adaptive Logics for Abduction.   Since abduction is a defeasible type of reasoning, adaptive logics are a good tool to model this type of relation.[3] The main advantages with respect to this problem can be summed up as follows.

Firstly, it allows for direct implementation of defeasible reasoning steps (*in casu* applications of *Affirming the Consequent*), which makes it possible to construct logical proofs that nicely integrate defeasible (ampliative in this case) and deductive inferences. This corresponds to the natural way in which humans reason.

Secondly, the formal apparatus of an adaptive logic instructs one to specify exactly which conditions would falsify the (defeasible) reasoning step. So, if this condition is derived later on in the proof, it defeats in a formal way all steps derived on the assumption that this condition is false. As these conditions are assumed—as long as one cannot derive them—to be false, they are called *abnormalities* in the adaptive logic literature. This possibility to defeat previous reasoning steps mirrors nicely the dynamics that is found in actual human reasoning.

Thirdly, there are for all adaptive logics in standard format generic proofs for most of the important metatheoretical properties (including soundness and completeness).

Within the adaptive logics program, several logics have been developed that model abductive reasoning. Practical abduction—in which the disjunction of the explanatory hypotheses is derived—is, for instance, adequately modelled in the logics $\mathbf{LA^r}$ and $\mathbf{LA^r_s}$.[4] Theoretical abduction has been modelled by the logic $\mathbf{AbL^t}$,[5] but here we will concentrate on the logic $\mathbf{MLA^s}$, developed in Gauderis (2011) for the purpose of modelling abductive reasoning in science. This logic provides an elegant way out of this problem by adding modalities to the object language and deriving—in case of our toy example that we used to introduce the problem—the hypotheses $\lozenge Qa$ and $\lozenge Ra$. In this way, the scientist or detective can work further on the individual hypotheses without having to prevent the conjunction, because $(\lozenge Qa \wedge \lozenge Ra)$ does not imply $\lozenge(Qa \wedge Ra)$ in any standard modal logic. The (Kripke-)semantics of this logic also foresees that everything that follows from investigating further one of the hypotheses is only verified within the possible world made accessible by formulating this hypothesis. The graphic representation of the semantics of our toy example below, illustrates that within this logic, different hypotheses and their conclusions are considered independently from each other.

---

[3]Some features of adaptive logics will be explained in order to make the paper understandable for people not familiar with adaptive logics. But the remarks made in this paper can, due to space limitations, hardly be called a good introduction to adaptive logics. We refer readers interested in adaptive logics to Batens (2011); Batens (2007) for a systematic overview and to Batens (2004) for a more philosophical defense of the use of adaptive logics.

[4]See Meheus and Batens (2006); Meheus (2007); Meheus (2011).

[5]See Lycke (2009).

$$w_0$$

$$\Box Pa$$

$$\Box(\forall x)(Qx \supset Px)$$

$$\Box(\forall x)(Rx \supset Px)$$

$$\Diamond Qa$$

$$\Diamond Ra$$

Aw_0w_1 Aw_0w_2

$$w_1 \qquad\qquad\qquad\qquad\qquad\qquad w_2$$

w_1	w_2
Qa	Ra
Pa	Pa
$(\forall x)(Qx \supset Px)$	$(\forall x)(Qx \supset Px)$
$(\forall x)(Rx \supset Px)$	$(\forall x)(Rx \supset Px)$

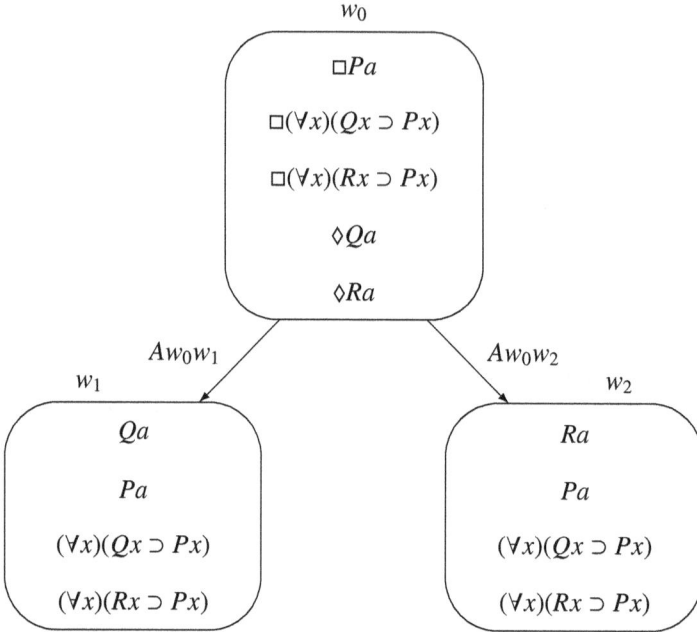

This feature makes the logic **MLA^s** (Modal Logic for Abduction) very apt to model, apart from its applications within science, a lot of everyday reasoning. The goal of this paper is to illustrate this with a crime investigation example. But before we present a more elaborate example, we will in the following section explain the logic **MLA^s**.

2 Formal Presentation of the Logic MLA^s

As any adaptive logic in the standard format, **MLA^s** is defined by a triple of a lower limit logic, a set of abnormalities and a strategy. These will be introduced in the following paragraphs after we have specified the language schema.

Formal Language Schema. Let $\mathcal{L}$ be the standard predicate language of **CL** with logical symbols $\neg, \supset, \wedge, \vee, \equiv, \forall$ and $\exists$. We will further use $\mathcal{C}, \mathcal{V}, \mathcal{F}$ and $\mathcal{W}$ to refer respectively to the sets of individual constants, individual variables, all (well-formed) formulas and the closed (well-formed) formulas of $\mathcal{L}$.

$\mathcal{L}_M$, the language of our logic, is $\mathcal{L}$ extended with the modal operators $\Box$ and $\Diamond$, where $\Box$ is primitive and $\Diamond$ defined in the usual way. $\mathcal{W}_M$, the set of closed formulas of $\mathcal{L}_M$ is the smallest set that satisfies the following conditions:

1. if $A \in \mathcal{W}$, then $A, \Box A, \Diamond A \in \mathcal{W}_M$

2. if $A \in \mathcal{W}_M$, then $\neg A \in \mathcal{W}_M$

3. if $A, B \in \mathcal{W}_M$, then $A \wedge B, A \vee B, A \supset B, A \equiv B \in \mathcal{W}_M$

It is important to note that there are—among other things—no occurrences of modal operators within the scope of another modal operator or a quantifier. We further define the set W_Γ—the subset of W_M, the elements of which can act as premises in our logic—as the smallest set that satisfies the following conditions:

1. if $A \in W$, then $\Box A, \Diamond A \in W_\Gamma$

2. if $A, B \in W_\Gamma$, then $A \wedge B \in W_\Gamma$

It is easily seen that $W_\Gamma \subset W_M$.

Lower Limit Logic.   Each adaptive logic is built on the deductive frame of a Tarski-logic. This lower limit logic (**LLL**) defines the undefeasible part of our logic. Every-thing that follows from the premises by the **LLL** will never be defeated.

The **LLL** of **MLA**^{s} is the predicate version of **D**, restricted by the language schema. **D** is characterized by a full axiomatization of predicate **CL** together with two axioms and an inference rule:

K $\Box(A \supset B) \supset (\Box A \supset \Box B)$

D $\Box A \supset \Diamond A$

NEC if $\vdash A$, then $\vdash \Box A$

The semantics for this logic can be expressed by a standard possible world Kripke semantics where the accessability relation A between possible worlds is *serial*, i.e. for every world w in our model, there is at least one world w' in our model such that wAw'. Soundness and completeness for **D** is—as for all normal modal logics—a well-established fact.

Set of Abnormalities.   The defeasible part of our logic is defined by the combination of the strategy and the set of abnormalities. This is a set of **LLL**-contingent formulas characterized by a logical form (or a union of such sets) that are assumed to be false 'as much as possible'. These assumptions allow us to derive, apart from the deductive consequences of the **LLL**, defeasible consequences that can be derived on a condi-tion, viz. the falsehood of the abnormalities. The inference rules are reduced to three generic rules: a premise, an unconditional and a conditional inference rule. An extra element is added on each line which is the set of conditions on which the formula on that line is derived.

PREM   If $A \in \Gamma$:

$$\frac{\vdots \qquad \vdots}{A \qquad \emptyset}$$

RU   If $A_1, ..., A_n \vdash_{\mathbf{LLL}} B$:

$$\begin{array}{cc} A_1 & \Delta_1 \\ \vdots & \vdots \\ A_n & \Delta_n \\ \hline B & \Delta_1 \cup ... \cup \Delta_n \end{array}$$

RC   If $A_1, ..., A_n \vdash_{\mathbf{LLL}} B \vee Dab(\Theta)$

$$\begin{array}{cc} A_1 & \Delta_1 \\ \vdots & \vdots \\ A_n & \Delta_n \\ \hline B & \Delta_1 \cup ... \cup \Delta_n \cup \Theta \end{array}$$

To define the set of abnormalities of **MLA**^{s},[6] we first need to introduce a new notation. Suppose that $A_{PCNF}(\alpha)$ is the *Prenex Conjunctive Normal Form* of $A(\alpha)$ and that for $Q_i \in \{\forall, \exists\}, \gamma_i \in \mathcal{V}, A_i(\alpha) \in \mathcal{F}$:

$$A_{PCNF}(\alpha) = (Q_1\gamma_1)\ldots(Q_m\gamma_m)(A_1(\alpha) \wedge \ldots \wedge A_n(\alpha)).$$

Then we can define $A_i^{-1}(\alpha)$ $(1 \leqslant i \leqslant n)$ as follows:

$$\text{if } n > 1 \quad : \quad A_i^{-1}(\alpha) =_{df} (Q_1\gamma_1)\ldots(Q_m\gamma_m)(A_1(\alpha) \wedge \ldots \wedge A_{i-1}(\alpha) \wedge$$
$$A_{i+1}(\alpha) \wedge \ldots \wedge A_n(\alpha)),$$
$$\text{if } n = 1 \quad : \quad A_1^{-1}(\alpha) =_{df} \top.$$

The idea is to have a notation for the formula formed by leaving out the ith conjunct. With this notation the set of abnormalities is defined as follows:

$$\Omega \quad = \quad \{\Box((\forall\alpha)(A(\alpha) \supset B(\alpha)) \wedge (B(\beta) \wedge \neg A(\beta)))$$
$$\vee \Box(\forall\alpha)B(\alpha) \vee \bigvee_{i=1}^{n} \Box(\forall\alpha)(A_i^{-1}(\alpha) \supset B(\alpha)) \mid$$
$$\text{No predicate that occurs in } B \text{ occurs in } A,$$
$$\alpha \in \mathcal{V}, \beta \in C, A, B \in \mathcal{F}\}.$$

This form might look complex, but its functioning is quite straightforward. We actually just made a disjunction of three reasons why we stop considering $A(\beta)$ as a good explanatory hypothesis for the phenomenon $B(\beta)$. These three possible reasons are (i) when $\neg A(\beta)$ is the case, (ii) when $B(\beta)$ is a tautology (and obviously, does not need an explanatory hypothesis) or (iii) when $A(\beta)$ has a redundant part and is therefore not an adequate explanatory hypothesis.

From now on, we can unambiguously shorten this logical form of the abnormalities as $!A(\beta) \triangleright B(\beta)$ which could be read as "$A(\beta)$ is not a valid hypothesis for $B(\beta)$".

Adaptive Strategy. Finally, a strategy is needed to define how to interpret the idea 'false as much as possible' exactly. In that way, the strategy orders which defeasible reasoning steps should be marked.[7]

Definition 2.1 (Marking for the simple strategy). *Line i with condition Δ is marked for the simple strategy at stage s of a proof,[8] if the stage s contains a line of which an $A \in \Delta$ is the formula and $\emptyset$ the condition.*

---

[6] The set of abnormalities is, due to space limitations, presented here quite briefly. For a more elaborate explanation how this set came to be, we refer to Gauderis (2011) in which the logic was presented for the first time.

[7] In adaptive logics, traditionally the mark $\checkmark$ is used to indicate that a step is defeated.

[8] A *stage of a proof* is a sequence of lines and a proof is a chain of stages. Every proof starts off with an empty sequence (stage 0). Each time a line is added to the proof by applying one of the inference rules, the proof comes to its next stage, which is the sequence of lines written so far including the new line.

Definition 2.2. *A formula A is* derived from Γ at stage *s of a proof iff A is the formula of a line that is unmarked at stage s.*

Definition 2.3. *A formula A is* finally derived from Γ at stage *s of a proof iff A is derived at line i, line i is not marked at stage s and remains unmarked in every extension of the proof.*[9]

Definition 2.4 (Final Derivability). *For* $\Gamma \subset \mathcal{W}_\Gamma$: $\Gamma \vdash_{\mathbf{MLA}^s} A$ *(A is finally* $\mathbf{MLA}^s$-derivable from Γ) *iff A is finally derived in a* $\mathbf{MLA}^s$-*proof from* Γ.

3 Application of $\mathbf{MLA}^s$ in Detective Literature

To illustrate the functioning of this logic, we will make up an original story. This is because we do not want to let our example grow too complex (and thus, less illustrative).[10]

> On a certain morning, X is found murdered in mysterious circumstances. From our first investigations we are able to determine three suspects (Sx) a, b and c who could be the murderer (Mx). Confronting these suspects with the facts, only c is able to pull out a water tight alibi (Ax) for the moment of the murder, 10.30am. Further, at the crime scene we find two clues: some long blond hairs at the murder weapon and a receipt of the tailor delivered at 9.30am, both of which could not have belonged to the victim.

The whole of this data constitutes our background knowledge. Formalized we get our initial premise set Γ. The exact meaning of the predicates describing the clues is defined as follows:

B_1x	"x was in the possession of some long blond hairs at 10.30am"
T_1x	"x was in the possession of the tailor receipt at 10.30am"
T_2x	"x received the tailor receipt at 9.30am"

Then, we can start off our proof (which models our reasoning process):[11]

1	$\Box(\forall x)(Mx \supset Sx)$	-;PREM	$\emptyset$
2	$\Box(\forall x)(Ax \supset \neg Mx)$	-;PREM	$\emptyset$
3	$\Box(Sa \wedge Sb \wedge Sc)$	-;PREM	$\emptyset$
4	$\Box(\neg Aa \wedge \neg Ab \wedge Ac)$	-;PREM	$\emptyset$
5	$\Box(\forall x)(Mx \equiv B_1x)$	-;PREM	$\emptyset$
6	$\Box(\forall x)(Mx \equiv T_1x)$	-;PREM	$\emptyset$

Now, the logic abduces three possible hypotheses, but the third one is—as expected—directly marked.

---

[9]This definition is slightly different from the more general definition mentioned in Batens (2011) because, using the simple strategy, it is in our case not possible that a marked line becomes unmarked at a later stage of a proof.

[10]Obviously we also don't want to spoil any plot of a classic in the genre.

[11]The only formalization that might appear a bit odd is the first line that seems to state that the murderer is always a suspect. Still, this line is the correct formalization, since this logic models a reasoning process that leads to a murderer. Consider the following: if the murderer was never suspected, he also will never be caught. So the actual meaning of the predicate Mx is more epistemological (like the predicate Sx essentially also is): "It can be shown that x is the murderer".

7	$\Diamond Ma$	1,3;RC	$\{!Ma \triangleright Sa\}$	
8	$\Diamond Mb$	1,3;RC	$\{!Mb \triangleright Sb\}$	
9	$\Diamond Mc$	1,3;RC	$\{!Mc \triangleright Sc\}$	$\checkmark^{11}$
10	$\Box \neg Mc$	2,4;RU	$\emptyset$	
11	$!Mc \triangleright Sc$	1,3,10;RU	$\emptyset$	

Left with two suspects, we need to investigate further the information of
our clues. On the one hand, b actually has long blond hair and could be
the owner of the found hairs, while a has ginger hair and could therefore
not be the owner. But on the other hand, the tailor (who wasn't aware
that a murder had happened) assures us that nobody with blond hair has
entered his shop that morning, while several people with ginger hair did.
Still, we have no reason to doubt the fact that it was the murderer himself
that was that morning in the shop.[12] Puzzled by this new information, we
bring the suspects a final visit to confront them with the clues. But when
we ring at a's door, the door is opened by a blond woman who says to be
a's wife.

We can now add this extra information to our proof.[13]

12	$\Diamond B_1 a \wedge \Box B_1 b$	-;PREM	$\emptyset$
13	$\Diamond T_2 a \wedge \Box \neg T_2 b$	-;PREM	$\emptyset$
14	$\Box(\forall x)(T_2 x \equiv T_1 x)$	-;PREM	$\emptyset$

With these new data, our detective can at this point rule out one more suspect. So,
only one suspect is left. It is the only hypothesis we can derive that is compatible with
all the known data. Here one can see again that hypothesis formation out of a certain
background knowledge is closely related with compatibility.

7	$\Diamond Ma$	1,3;RC	$\{!Ma \triangleright Sa\}$	
8	$\Diamond Mb$	1,3;RC	$\{!Mb \triangleright Sb\}$	$\checkmark^{16}$
15	$\Box \neg Mb$	6,13,14;RU	$\emptyset$	
16	$!Mb \triangleright Sb$	1,3,15;RU	$\emptyset$	

Some Concluding Remarks Is it also actually possible to come with conclusive evidence? Here we have to keep in mind that we are dealing with a logic for abduction
or hypothesis formation. This should not be confused with deciding whether we have
conclusive evidence. In science too, the forming of hypotheses and the confirmation
of theories are two different steps in the scientific process. These two different methods were actually the two things that Charles Peirce wanted to keep seperate in his
methodology of science by discerning *abduction* and *induction*.

---

[12] This assumption is made because we don't want the example to grow too complex; but it actually also
nicely illustrates how many 'hidden assumptions' there are in a reasoning process, assumptions that have to
be spelled out fully in a formal logic.

[13] Technically speaking, we have a new premise set Γ' and need to start a new proof, but it is easily seen
that we can start our new proof by copying the previous proof and continue to add the new data as premises.

At the end of the day, we illustrated in this paper that the logic **MLA**^{s}, developed to model abduction in science, can also be used in more everyday life situations, as, for instance, our exploration in the detective genre shows. At first sight, the reasoning process seems quite equivalent. From a formal point of view, there is actually one major difference. A detective will point his attention to which subject is the murderer; his hypotheses are, for instance, *Ma*, *Mb* and *Mc*. A scientist on the other hand, will focus more on the predicates or properties; his hypotheses for a puzzling phenomenon *Pa* will most likely be modelled with hypotheses of the form *Fa*, *Ga* and *Ha*. It is an interesting route to investigate whether this formal difference leads to more differences between the two types of reasoning.

References

Batens, D. (2004). "The Need for Adaptive Logics in Epistemology". In: *Logic, Epistemology and the Unity of Science*. Ed. by D. Gabbay et al. Dordrecht: Kluwer Academic Publishers.

— (2007). "A Universal Logic Approach to Adaptive Logics". In: *Logica Universalis* 1, pp. 221–242.

— (2011). *Adaptive Logics and Dynamic Proofs. A Study in the Dynamics of Reasoning*. Forthcoming.

Gauderis, T. (2011). "Modelling Abduction in Science by Means of a Modal Adaptive logic". In: *Foundations of Science (forthcoming)*.

Lycke, H. (2009). "The Adaptive Logics Approach to Abduction". In: *Logic, Philosophy and History of Science in Belgium. Proceedings of the Young Researchers Days 2008*. Ed. by E. Weber et al. Brussel: Koninklijke Vlaamse Academie van België, pp. 35–41.

Meheus, J. (2007). "Adaptive Logics for Abduction and the Explication of Explanation-Seeking Processes." In: *Abduction and the Process of Scientific Discovery*. Ed. by O. Pombo and A. Gerne. Lisboa: Centro de Filosofia das Ciências da U. de Lisboa, pp. 97–119.

— (2011). "A Formal Logic for the Abductions of Singular Hypotheses". Forthcoming.

Meheus, J. and D. Batens (2006). "A Formal Logic for Abductive Reasoning". In: *Logic Journal of the IGPL* 14, pp. 221–236.

Peirce, C. S. (1958–60). *Collected Papers*. Cambridge, MA: Belknap Press of Harvard University Press.

Sebeok, T. A. and U. Eco, eds. (1988). *The Sign of Three: Dupin, Holmes and Peirce*. Bloomington, IN: Indiana University Press.

CHAPTER 6

Constructing the Lindenbaum Algebra for a Logic Step-by-Step Using Duality

Sam van Gool

Abstract

We discuss the incremental construction of the Lindenbaum algebra for a modal logic, using discrete duality for Boolean algebras with operators. We work out in detail the case of the modal logic T, as an illustrative example of the method for modal logics with mixed-rank axiomatizations.[1]

1 Introduction

In the study of a propositional logic $\mathcal{L}$, the following construction is often important: take the set of all formulas in the language of $\mathcal{L}$, and partition this set into classes of $\mathcal{L}$-equivalent formulas. In many cases, the set of $\mathcal{L}$-equivalence classes has a natural algebraic structure, which is called the *Lindenbaum algebra* for the logic $\mathcal{L}$.

Algebraic methods are useful to study issues such as term complexity, decidability of logical equivalence, interpolation, and normal forms for a logic, i.e., problems in which one considers formulas whose variables are drawn from a finite set. The oldest instance of the use of algebraic methods in logic goes back to George Boole: the Lindenbaum algebra for *classical propositional logic* (CPL) on n variables $\{p_1, \ldots, p_n\}$ is a *Boolean algebra*, and it can be shown to be (isomorphic to) $\mathcal{P}(\mathcal{P}(\{p_1, \ldots, p_n\}))$, which, as we will explain below, is the free Boolean algebra on n generators. The logical impact of this result is the disjunctive normal form theorem for CPL. However,

---

[1]The results described in this chapter were a first step towards the results in the paper Gool and Coumans (2012), joint with Dion Coumans.

for logics other than CPL, the situation is often much more complicated, and a simple description of the Lindenbaum algebra is usually not available. For example, the Lindenbaum algebra for *intuitionistic propositional logic* (IPL) on only two variables, i.e., the free Heyting algebra on two generators, is already infinite and non-trivial to describe.

Modal logics form another rich class of examples of logics whose Lindenbaum algebras are often infinite and complicated. These logics are based on CPL, enriched with a unary connective '◇', which is meant to formalize a notion of 'possibility'. Different axioms for ◇ yield different modal logics. One may try to gain a better understanding of a particular modal logic through its Lindenbaum algebra. As a representative example, we will mainly concentrate on the Lindenbaum algebra for a very simple modal logic called T, and we will also indicate how we expect the described methods may apply to a larger class of logics.

To motivate our methodology, we start from the following well-known definition of the language of propositional logic with one unary modality (see, for instance, Blackburn, de Rijke, and Venema (2001)).

Definition 1.1. *Let P be a set of* **propositional variables**. *The set of* **propositional formulas in P**, Prop(P), *is defined to be the smallest set which contains P, $\bot$, $\top$, and is closed under the formation rules: for all $\varphi, \psi \in$ Prop(P), $\varphi \wedge \psi$, $\varphi \vee \psi$, $\varphi \to \psi$ and $\neg \varphi$ are in* Prop(P).

The sets $\Phi_n(P)$ of **modal formulas in P of rank at most n** *are defined inductively as follows.*

$$\Phi_0(P) := \text{Prop}(P),$$
$$\Phi_{n+1}(P) := \text{Prop}(P \cup \{\Diamond\varphi : \varphi \in \Phi_n(P)\}).$$

The set of **modal formulas in P** *is defined as $\Phi(P) := \bigcup_{n\in\mathbb{N}} \Phi_n(P)$. In algebraic terms, $\Phi(P)$ is the domain of the* **absolutely free algebra** *in the signature $\{\bot,\top,\neg,\Diamond,\wedge,\vee,\to\}$.*

This *step-by-step* construction of the absolutely free algebra (i.e., the algebra describing the modal *language*) is a first example of the construction that we will also use for the Lindenbaum algebra (i.e., the algebra describing a modal *logic*). More precisely, we aim to understand the Lindenbaum algebra $\mathbb{B}$ for a modal logic $\mathcal{L}$ in a layered manner: for each $n \geq 0$, the $\mathcal{L}$-equivalence classes of modal formulas of rank at most n form a Boolean subalgebra $\mathbb{B}_n$ of $\mathbb{B}$. We thus get a chain of Boolean algebras

$$\mathbb{B}_0 \rightarrowtail \mathbb{B}_1 \rightarrowtail \mathbb{B}_2 \rightarrowtail \cdots ,$$

and the Lindenbaum algebra is then the direct limit, or colimit, of this chain (see Section 3.4 below). Each of the algebras $\mathbb{B}_n$ will be finite, and $\mathbb{B}_n$ embeds into $\mathbb{B}_{n+1}$ for each n. These two properties imply that the chain accurately approximates the infinite Lindenbaum algebra $\mathbb{B}$ by its finite pieces.

In certain well-behaved examples, the finite pieces $\mathbb{B}_0$, $\mathbb{B}_1$, ... of the Lindenbaum algebra can be described by a uniform process of the following kind: start from a simple $\mathbb{B}_0$ (usually the Lindenbaum algebra for CPL), and then define $\mathbb{B}_{n+1}$ from $\mathbb{B}_n$ in a uniform way. As an immediate application, one then obtains an algorithm for deciding $\mathcal{L}$-equivalence: given two formulas φ, ψ, let n be the maximum of the ranks of φ and ψ (a formal definition of *rank* will be given below), and check whether φ and

ψ are equal under all interpretations of the propositional variables in the finite algebra $\mathbb{B}_n$. Examples of logics for which the step-by-step construction works are (trivially) CPL, but also the basic modal logic K, the modal logic S4, which is closely connected to IPL, and, as we will show in this paper, the modal logic T.

All these examples now beg an interesting question, which one may ask for any given modal logic $\mathcal{L}$:

Is there a uniform step-by-step construction of the Lindenbaum algebra for $\mathcal{L}$?

$$(Q_{\mathcal{L}})$$

With the syntactic definition of *rank*, we can describe an important class of examples of logics for which the answer to the question $(Q_{\mathcal{L}})$ is affirmative.

For a modal formula $\varphi \in \Phi(P)$, the **rank** of an occurrence of a propositional variable $p \in P$ is the number of $\diamond$'s having that occurrence of p in its scope. We say that φ **is of rank at most** n if every occurrence of a variable in φ is of rank less than or equal to n, and that φ **is of rank exactly** n or **pure rank** n if every occurrence of a variable in the formula is of rank equal to n. We also say, for instance, "φ is of **(mixed) rank 1-2**" if every occurrence of a variable is of rank either 1 or 2.

For example, the formulas $\diamond(p \vee q)$ and $\diamond p \vee \diamond q$ are of rank exactly 1, whereas the formula $\diamond p \wedge p$ is not of rank exactly 1, but it is of rank 0-1. The formula $\diamond\diamond p \wedge \diamond p$ is of rank 1-2.

If all axioms for a logic $\mathcal{L}$ are of rank exactly 1, then the answer to the question $(Q_{\mathcal{L}})$ can be shown to be affirmative, using the fact that $\mathcal{L}$ is in the realm of 'algebras for a functor' (cf. Bonsangue and Kurz 2006 and Bezhanishvili and Kurz 2007 for detailed accounts). As a result, one directly obtains a normal form theorem and an algorithm for deciding $\mathcal{L}$-equivalence for all rank 1 logics.

On the other side of the spectrum, there exist modal logics in which $\mathcal{L}$-equivalence is undecidable (cf. Chapter 6 of Blackburn, de Rijke, and Venema 2001), so that, in particular, one cannot hope to have an affirmative answer to the question $(Q_{\mathcal{L}})$ for such logics. As far as we know, the most general class for which a positive answer to the question $(Q_{\mathcal{L}})$ has been given is the class of logics with axioms of rank exactly 1. However, Ghilardi (1992), Ghilardi (2010) and Bezhanishvili and Gehrke (2011) have widened the scope of the uniform step-by-step approach to a few particular logics that are not axiomatizable by pure rank 1 axioms, namely IPL and S4. In this paper, we will discuss another example of a rank 0-1 logic which admits a step-by-step approach, namely the modal logic T, the modal logic of reflexive frames. For the line of research outlined above, it is important to properly understand the example of the modal logic T, because it falls 'in between' the basic modal logic K and the logic S4, so that it is on the one hand simpler than S4, but on the other hand already needs much of the same machinery as is needed for S4, and falls outside the realm of pure rank 1 logics. In the final section of Bezhanishvili and Kurz (2007), the authors already briefly indicate how to see the modal logic T as a first step outside pure rank 1 axioms. We will work this out in detail here.

A key feature of the approach to answering the question $(Q_{\mathcal{L}})$ is the use of *duality* to obtain a concrete, set-theoretic, *dual description* of the finite algebras in the approximating sequence. Important questions about the chain of algebras and homomorphisms can then be answered more quickly by looking at the dual chain of sets and functions. Most crucially, for the step-by-step construction of the Lindenbaum algebra to be useful in applications, one needs the homomorphisms $\mathbb{B}_n \to \mathbb{B}_{n+1}$ in the chain to be injective. Using the dual description, this amounts to checking that certain functions are surjective (cf. Proposition 3.3 below). One thus reduces an abstract, formulaic question about algebra to a concrete, spatial question about sets. As such, this approach is a typical example of the use of duality in algebra and logic.

The outline of this paper is as follows. We will start by reviewing some preliminaries on universal algebra and on duality for finite Boolean algebras in Section 2, which we will need, respectively, to justify the step-by-step construction of the Lindenbaum algebra, and its dual description. We will discuss the construction and its dual description in Section 3, which forms the heart of this paper. Finally, we will point at some possible applications and future questions in our concluding Section 4.

2 Preliminaries

2.1 THE LINDENBAUM ALGEBRA AS A FREE ALGEBRA FOR A VARIETY

In this subsection, we quickly review the necessary background from universal algebra which justifies the step-by-step construction of the Lindenbaum algebra. We will not go into much detail, and will assume a basic understanding of the algebraic perspective on logic, cf. Blackburn, de Rijke, and Venema (2001) for more details.

From the perspective of algebraic logic, a logic which can be defined by equations naturally gives rise to an equationally defined class of algebras, i.e., a *variety*. For example, CPL corresponds to the variety of Boolean algebras, and IPL corresponds to the variety of Heyting algebras. The basic modal logic K is defined by adding an operator $\Diamond$ to the language of CPL (as in Definition 1.1 above), and adding as axioms the logical equivalences

$$\Diamond \bot \leftrightarrow \bot \tag{1}$$
$$\Diamond(a \lor b) \leftrightarrow \Diamond a \lor \Diamond b. \tag{2}$$

Modal logics also naturally correspond to varieties of algebras, which are called 'modal algebras'.

Definition 2.1. *A* **modal algebra** *is a pair* $(\mathbb{B}, \Diamond^{\mathbb{B}})$*, where* $\mathbb{B}$ *is a Boolean algebra, and* $\Diamond^{\mathbb{B}} \colon \mathbb{B} \to \mathbb{B}$ *is a unary operation preserving* $\bot$ *and binary joins. A* **homomorphism** $f \colon (\mathbb{B}, \Diamond^{\mathbb{B}}) \to (\mathbb{C}, \Diamond^{\mathbb{C}})$ *of modal algebras is a Boolean algebra homomorphism* $f \colon \mathbb{B} \to \mathbb{C}$ *such that* $f \circ \Diamond^{\mathbb{B}} = \Diamond^{\mathbb{C}} \circ f$.

Let $\mathcal{L}$ *be a modal logic which contains the basic modal logic* K. *The* **variety of** $\mathcal{L}$-**algebras,** $\mathcal{V}_{\mathcal{L}}$, *is defined to be the class of modal algebras in which all* $\mathcal{L}$-*equivalent formulas are equal under all interpretations of the variables. In particular,* $\mathcal{V}_{\mathsf{K}}$ *is the variety of all modal algebras.*

Recall that we defined the Lindenbaum algebra for a logic $\mathcal{L}$ as the quotient of the absolutely free algebra $\Phi(X)$ by the relation of $\mathcal{L}$-equivalence. The important fact from universal algebra that we need is that the Lindenbaum algebra for the logic $\mathcal{L}$ is exactly the so-called **free algebra for the variety** $\mathcal{V}_{\mathcal{L}}$.

Proposition 2.2. *Let* $\mathbb{B}$ *be the Lindenbaum algebra on a set of propositional variables P for a modal logic* $\mathcal{L}$. *Then* $\mathbb{B}$ *is the free* $\mathcal{V}_{\mathcal{L}}$-*algebra over P, i.e.,* $\mathbb{B}$ *is a P-generated algebra in* $\mathcal{V}_{\mathcal{L}}$ *with the following* **universal mapping property**: *for any* $\mathbb{A} \in \mathcal{V}_{\mathcal{L}}$ *and any function* $f \colon P \to \mathbb{A}$*, there exists a (necessarily unique) modal algebra homomorphism* $\overline{f} \colon \mathbb{B} \to \mathbb{A}$ *which extends f.*

Proof. Cf. any reference on universal algebra or algebraic logic, for example Burris and Sankappanavar (2000), Theorem 10.10. □

For us, the importance of this Proposition is that in order to construct the Lindenbaum algebra for a logic $\mathcal{L}$ on n propositional variables, we may now construct an algebra $\mathbb{B}$ and show that it is the free $\mathcal{V}_{\mathcal{L}}$-algebra on n generators. As we pointed out in the introduction, logical problems, such as deciding logical equivalence, then become equivalent to algebraic problems, i.e., deciding equality in the algebra $\mathbb{B}$.

2.2 DUALITY FOR FINITE BOOLEAN ALGEBRAS

The dual, set-theoretic, description of the chain of Boolean algebras that we will build relies on so-called discrete duality for finite Boolean algebras. The results in this section have been around since Stone (1936), and there is no claim of originality. Therefore, and also for the sake of readability and brevity, we will skip some of the proofs, and instead refer to the extended online version of this paper (Coumans and Gool 2011) for detailed proofs.

Let us first introduce the following notation for the categories that are involved.

Category	Objects	Morphisms
$\mathsf{BA}_{<\omega}$	finite Boolean algebras	Boolean algebra homomorphisms
$\mathsf{BA}^{\vee}_{<\omega}$	finite Boolean algebras	$\vee$-semilattice homomorphisms
$\mathsf{Set}_{<\omega}$	finite sets	functions
$\mathsf{Set}^{\mathsf{Rel}}_{<\omega}$	finite sets	relations

The following is the basic Stone duality result in the finite case. We will sketch the proof and, while doing so, introduce notation that we will use later.

Theorem 2.3. *The categories $\mathsf{BA}^{\vee}_{<\omega}$ and $\mathsf{Set}^{\mathsf{Rel}}_{<\omega}$ are dually equivalent. The dual equivalence restricts to a dual equivalence between the categories $\mathsf{BA}_{<\omega}$ and $\mathsf{Set}_{<\omega}$.*

Proof (Sketch). Given a finite set X, the *power set algebra* $\mathbb{P}(X)$ is the Boolean algebra based on the power set $\mathcal{P}(X)$ of X. Given a Boolean algebra $\mathbb{B}$, its dual is defined to be the *set of atoms*[2] of $\mathbb{B}$, which we denote by $\mathsf{At}(\mathbb{B})$. Given a relation $R\colon X \to Y$ between finite sets, we have a Boolean algebra hemimorphism $\Diamond_R\colon \mathbb{P}(Y) \to \mathbb{P}(X)$, defined by $\Diamond_R(U) := \{x \in X \mid \exists y \in U : xRy\}$. Given a join-preserving function $h\colon \mathbb{B} \to \mathbb{C}$ between finite Boolean algebras, we define a relation $R_h\colon \mathsf{At}(\mathbb{C}) \to \mathsf{At}(\mathbb{B})$ by $x_C\, R_h\, x_B \iff x_C \leq h(x_B)$. The relation R_h is functional iff h is also meet-preserving. See the extended version of this paper for more details. □

There are several consequences of the dual equivalence $\mathsf{BA}_{<\omega} \leftrightarrows \mathsf{Set}_{<\omega}$ that we will often use. Firstly, it allows us to describe coproducts of Boolean algebras using products of sets, as follows.

Proposition 2.4. *Let $\mathbb{B}$ and $\mathbb{C}$ be finite Boolean algebras. The coproduct $\mathbb{B} + \mathbb{C}$ is isomorphic to $\mathbb{P}(\mathsf{At}(\mathbb{B}) \times \mathsf{At}(\mathbb{C}))$. The coproduct injection $\mathbb{B} \to \mathbb{B} + \mathbb{C}$ corresponds under the duality to the projection function $\mathsf{At}(\mathbb{B}) \times \mathsf{At}(\mathbb{C}) \to \mathsf{At}(\mathbb{B})$, and similarly for the injection $\mathbb{C} \to \mathbb{B} + \mathbb{C}$.*

---

[2] An **atom** of $\mathbb{B}$ is a non-zero element $a \in \mathbb{B}$ which does not have any non-zero elements below it.

Proof. See the extended version of this paper, Coumans and Gool (2011). □

The second consequence is that a surjective homomorphism (i.e., epimorphism) $\pi: \mathbb{B} \to \mathbb{C}$ of finite Boolean algebras corresponds to an injective function (i.e., monomorphism) $i_\pi: \text{At}(\mathbb{C}) \to \text{At}(\mathbb{B})$ of sets. So *the atoms of a quotient of* $\mathbb{B}$ *may be identified with a subset of the atoms of* $\mathbb{B}$. We can be a bit more specific, introducing some notation: if $h: \mathbb{B} \to \mathbb{C}$ is a BA homomorphism, recall that its *kernel* is the set $\ker(h) := \{(b, b') \in B^2 : h(b) = h(b')\}$, and that atoms $x_B \in \text{At}(\mathbb{B})$ may be identified with homomorphisms $h_{x_B}: \mathbb{B} \to \mathbf{2}$.

Proposition 2.5. *Let $\pi: \mathbb{B} \to \mathbb{C}$ be a surjective homomorphism, and $i_\pi: \text{At}(\mathbb{C}) \to \text{At}(\mathbb{B})$ the injective function dual to π. Then*

$$i_\pi[\text{At}(\mathbb{C})] = \{x_B \in \text{At}(\mathbb{B}) : \ker(\pi) \subseteq \ker(h_{x_B})\}$$
$$= \{x_B \in \text{At}(\mathbb{B}) : \forall b, b' \in B : \pi(b) = \pi(b') \to (x_B \leq b \leftrightarrow x_B \leq b')\}.$$

Proof. See the extended version of this paper. □

The duality can also be used to give concrete descriptions of the free Boolean algebra over P, $F_{\text{BA}}(P)$.

Proposition 2.6. *There is a bijective function between the sets $\text{At}(F_{\text{BA}}(P))$ and $\mathcal{P}(P)$. It can be given explicitly by sending an atom x of $F_{\text{BA}}(P)$ to the set $\{p \in P : x \leq p\}$, and, conversely, sending $A \subseteq P$ to $z(A) := \bigwedge_{p \in A} p \wedge \bigwedge_{q \notin A} \neg q$.*

Proof. See the extended version of this paper. □

3 Construction of the Free T-algebra

We will concentrate on the construction of the finitely generated free algebras for the variety $\mathcal{V}_\mathsf{T} \subseteq \mathcal{V}_\mathsf{K}$, consisting of the modal algebras $(\mathbb{B}, \Diamond)$ satisfying the additional equation

$$a \leq \Diamond a. \tag{T}$$

(Note that any inequality is indeed equivalent to an equation, since in any lattice we have $a \leq b$ iff $a \wedge b = a$.)

3.1 A CHAIN OF FINITE BOOLEAN ALGEBRAS

Let P be a finite set of proposition letters. We are going to construct a chain diagram in the category of Boolean algebras, whose colimit will be the Boolean algebra underlying the free modal algebra for $\mathcal{V}_\mathsf{T}$ over P.

Let $\mathbb{B}_0 := F_{\text{BA}}(P)$, the free Boolean algebra over P.

For $n \geq 0$, assume $\mathbb{B}_n$ has been defined by induction. Denote by $\blacklozenge B_n$ the *set* of symbols $\{\blacklozenge b : b \in B_n\}$. Let $\mathbb{C}_{n+1} := \mathbb{B}_n + F_{\text{BA}}(\blacklozenge B_n)$. We emphasize that in the second summand, we are taking the free Boolean algebra over the *set* $\blacklozenge B_n$ underlying the Boolean algebra $\mathbb{B}_n$, forgetting all the structure of $\mathbb{B}_n$ for this part.

Denote the coproduct injection of the first coordinate by $j_n: \mathbb{B}_n \to \mathbb{C}_{n+1}$, which is a BA homomorphism. Further let $\blacklozenge_n: B_n \to C_{n+1}$ be the composition of the universal arrow $B_n \to F_{BA}(\blacklozenge B_n)$, followed by coproduct injection $F_{BA}(\blacklozenge B_n) \to \mathbb{C}_{n+1}$ of the second coordinate. We emphasize again that, so far, $\blacklozenge_n$ is only a function between the domains of the algebras.

Consider the following equations, for $a_n, b_n \in B_n$, $a_{n-1} \in B_{n-1}$:

$$\blacklozenge \bot = \bot \tag{1_n}$$

$$\blacklozenge (a_n \vee b_n) = \blacklozenge a_n \vee \blacklozenge b_n \tag{2_n}$$

$$j_n a_n \leq \blacklozenge a_n \tag{T_n}$$

$$j_n \blacklozenge_{n-1} a_{n-1} = \blacklozenge i_{n-1} a_{n-1} \tag{W_n}$$

Let us write $\mathcal{E}_n$ for the set of instances of the equations (1_n), (2_n), (T_n), (W_n). We may view $\mathcal{E}_n$ as the set of those pairs $(c_{n+1}, d_{n+1}) \in \mathbb{C}_{n+1} \times \mathbb{C}_{n+1}$ such that '$c_{n+1} = d_{n+1}$' is an instance of one of these equations.[3] Then $\mathcal{E}_n$ generates a congruence relation $\approx_n$ on $\mathbb{C}_{n+1}$. We define $\mathbb{B}_{n+1}$ to be the quotient $\mathbb{C}_{n+1}/\approx_n$ and $p_n \colon \mathbb{C}_{n+1} \to \mathbb{B}_{n+1}$ the canonical projection, for which $\ker(p_n) = \approx_n$. We define $i_n := p_n \circ j_n \colon B_n \to B_{n+1}$, and, by a slight abuse of notation, also denote the function $p_n \circ \blacklozenge_n$ by $\blacklozenge_n$. Note that i_n is a BA homomorphism, and moreover that $\blacklozenge_n$ is now a join-preserving function, because equations (1_n) and (2_n) hold in $\mathbb{B}_{n+1}$. Intuitively, the axiom (T_n) is included to ensure that we are really building an algebra in the variety $\mathcal{V}_T$. Axiom (W_n) is needed so that we will end up having a well-defined operator $\Diamond$ on the colimit of the chain of Boolean algebras. This argument will be made precise in Subsection 3.4 below.

We thus get the following chains of (finite) Boolean algebras:

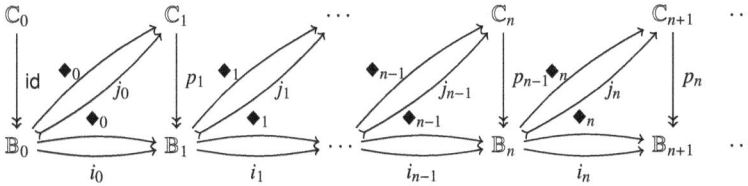

3.2 THE DUAL CHAIN OF FINITE SETS

We will now explicitly calculate what the sets, functions and relations dual to the Boolean algebras and morphisms in the above-defined chain are.

Since $\mathbb{B}_0 = F_{BA}(P)$, the set X_0 can be identified with $\mathcal{P}(P)$, by Proposition 2.6. In the rest of this subsection, we will calculate an explicit description of the set $X_{n+1} = \mathrm{At}(\mathbb{B}_{n+1})$, assuming the sets X_m for $m \leq n$ have been described.

First of all, we note that $\mathrm{At}(\mathbb{C}_{n+1})$ is equal to $X_n \times \mathcal{P}(B_n)$, using Propositions 2.4 and 2.6. We write an arbitrary atom of $\mathbb{C}_{n+1}$ as $x_C = (x, A)$, where $x \in X_n$ and $A \subseteq B_n$.

By Proposition 2.4, the dual of $j_n \colon B_n \to B_n + F_{BA}(\blacklozenge B_n)$ is given by the projection onto the first coordinate, $\rho_n \colon X_n \times \mathcal{P}(B_n) \to X_n$. Its restriction to X_{n+1} is dual to the map $i_n \colon B_n \to B_{n+1}$, and we will denote it by π_n.

Summing up, we now have the following diagram of finite sets, functions and relations, the relation $R_{\blacklozenge_n}$ being dual to the join-preserving function $\blacklozenge_n \colon B_n \to B_{n+1}$, recalling Theorem 2.3.

---

[3] For the equation (W_n) to make sense when $n = 0$, we put $\mathbb{B}_{-1} := \mathbf{2}$, and $i_{-1} = \blacklozenge_{-1}$ the homomorphism $\mathbb{B}_{-1} \to \mathbb{B}_0$, and note that (W_n) then only says $\blacklozenge_0 \top = \top$ and $\blacklozenge_0 \bot = \bot$, which already follows from equations (1_0) and (T_0).

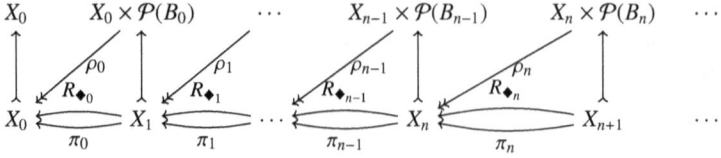

The following lemma will give concrete conditions under which certain inequalities in $\mathbb{C}_{n+1}$ hold.

Lemma 3.1. *Let $\mathbb{B}$ be a Boolean algebra, X its set of atoms, $\mathbb{C} := \mathbb{B} + F_{\mathsf{BA}}(\blacklozenge\mathbb{B})$, and j and $\blacklozenge$ the obvious maps $\mathbb{B} \to \mathbb{C}$. For any atom $(x, A) \in X \times \mathcal{P}(B)$ of $\mathbb{C}$ and $b \in \mathbb{B}$, we have*

1. *$(x, A) \le j(b)$ in $\mathbb{C} \iff x \le b$ in $\mathbb{B}$,*

2. *$(x, A) \le \blacklozenge b$ in $\mathbb{C} \iff b \in A$.*

Proof. See the extended version of this paper. $\square$

Now, by Proposition 2.5, the quotient $\mathbb{B}_{n+1}$ of $\mathbb{C}_{n+1}$ has as its dual the set of those atoms $x_C \in \mathrm{At}(\mathbb{C}_{n+1})$ for which $\approx_n$ is contained in $\ker(x_C)$. Concretely, this happens if and only if the atom x_C 'satisfies' all the inequalities in $\mathcal{E}_n$. Here, we say an atom x_C *satisfies* the formal inequality '$s(x_1, \ldots, x_i) \le t(x_1, \ldots, x_i)$' if, for all $a_1, \ldots, a_i \in \mathbb{B}_n$, we have $x_C \le s(a_1, \ldots, a_i)$ implies $x_C \le t(a_1, \ldots, a_i)$.

Let (x, A) be an atom of $\mathbb{C}_{n+1} = \mathbb{B}_n + F_{\mathsf{BA}}(\blacklozenge\mathbb{B}_n)$. We will repeatedly apply Lemma 3.1 to give conditions under which (x, A) satisfies the equations in $\mathcal{E}_n$.

- $(1_n) : \blacklozenge\bot = \bot$.

  Since we have $(x, A) \not\le \bot$, we need that $(x, A) \not\le \blacklozenge\bot$. By Lemma 3.1, this happens iff

  $$\bot \notin A. \qquad (1_n^{\partial})$$

- $(2_n) : \blacklozenge(a \vee b) = \blacklozenge a \vee \blacklozenge b$.

  We need to have $(x, A) \le \blacklozenge(a \vee b)$ iff $(x, A) \le \blacklozenge a \vee \blacklozenge b$. Since (x, A) is an atom, we have $(x, A) \le \blacklozenge a \vee \blacklozenge b$ iff $(x, A) \le \blacklozenge a$ or $(x, A) \le \blacklozenge b$, which happens iff $a \in A$ or $b \in A$. We thus get the following condition on A:

  $$\forall a, b \in \mathbb{B}_n : a \vee b \in A \iff a \in A \text{ or } b \in A. \qquad (2_n^{\partial})$$

- $(T_n) : j_n a \le \blacklozenge a$.

  Using Lemma 3.1 again, we have $(x, A) \le j_n a$ iff $x \le a$ and $(x, A) \le \blacklozenge a$ iff $a \in A$. We thus get the condition

  $$\forall a \in \mathbb{B}_n : x \le a \Rightarrow a \in A. \qquad (T_n^{\partial})$$

- $(W_n) : j_n \blacklozenge_{n-1} a_{n-1} = \blacklozenge i_{n-1} a_{n-1}$

  One may calculate[4] that $((x, A), \mathcal{A})$ satisfies (W_n) iff

  $$\forall y_{n-1} \in \mathrm{At}(\mathbb{B}_{n-1})[(\exists B \subseteq \mathbb{B}_{n-1} : (y_{n-1}, B) \in \mathcal{A}) \iff y_{n-1} \in A],$$

---

[4] Again, we refer the reader to the extended online version of this paper for detailed calculations.

or, more concisely, iff

$$\rho_{n-1}[\mathcal{A}] = A \cap \mathsf{At}(\mathbb{B}_{n-1}), \qquad\qquad (W_n^\partial)$$

where ρ_{n-1} is the projection $X_{n-1} \times \mathcal{P}(B_{n-1}) \to X_{n-1}$ onto the first coordinate.

Now that we have the conditions (1_n^∂), (2_n^∂), (T_n^∂) and (W_n^∂) for the atom $x_C = (x, A) \in \mathsf{At}(\mathbb{C}_{n+1})$ to be an atom of $\mathbb{B}_{n+1}$, we can give a more concise description.

Lemma 3.2. *Let $A \subseteq \mathbb{B}_n$. Then A satisfies (1_n^∂) and (2_n^∂) iff $A = \uparrow(A \cap \mathsf{At}(\mathbb{B}_n))$.*

Proof. See the extended version of this paper. □

Note that condition (T_n^∂) always implies that $x \in A$, but in case A is an upset, the condition '$x \in A$' is even *equivalent* to condition (T_n^∂). We conclude that the pairs (x, A) satisfying (1_n^∂), (2_n^∂) and (T_n^∂) are exactly those for which $A = \uparrow(A \cap \mathsf{At}(\mathbb{B}_n))$ and $x \in A$. Thus, by sending (x, A) to $(x, A \cap \mathsf{At}(\mathbb{B}_n))$, we may identify the atoms of $\mathbb{B}_{n+1}$ with certain pairs (x, T), where $T \subseteq \mathsf{At}(\mathbb{B}_n) = X_n$ and $x \in T$.

For $n = 0$, in fact all such pairs are atoms of $\mathbb{B}_1$, since the equation (W_0) is vacuously true. So we get

$$X_1 \cong \{(x, T) : x \in X_0, T \subseteq X_0, x \in T\}. \qquad\qquad (6.1)$$

For $n \geq 1$, we need to restrict to those atoms of $\mathbb{C}_{n+1}$ which also satisfy (W_n^∂). We then get

$$X_{n+1} \cong \{((x, T), \mathcal{A}) : (x, T) \in X_n, \mathcal{A} \subseteq X_n, (x, T) \in \mathcal{A}, T = \rho_{n-1}[\mathcal{A}]\}. \qquad (6.2)$$

We can now also calculate what the dual of the join-preserving function $\blacklozenge_n : B_n \to B_{n+1}$ is. By Stone duality, it will be the relation $R_{\blacklozenge_n} : X_{n+1} \to X_n$, given by

$$(x, T) R_{\blacklozenge_n} x' \iff (x, T) \leq \blacklozenge_n x'.$$

Applying Lemma 3.1 to the set $A = \uparrow T$ and using the fact that the atoms of $\mathbb{B}_n$ form an antichain, we now easily see that $(x, T) R_{\blacklozenge_n} x'$ iff $x' \in T$.

3.3 APPLICATION OF DUALITY: INJECTIVITY OF THE CHAIN

We can use the dual description of the chain of Boolean algebras to give an easy proof of the following fact.

Proposition 3.3. *For each n, the map $i_n : \mathbb{B}_n \to \mathbb{B}_{n+1}$ is injective.*

Proof. The statement is, by duality, equivalent to: for each n, the map $\pi_n : X_{n+1} \to X_n$ is surjective. We will prove this by induction. Recall that π_n is the restriction of the projection map $X_n \times \mathcal{P}(B_n) \to X_n$ to X_{n+1}. As $X_1 = \{(x, T) \in X_0 \times \mathcal{P}(X_0) \,|\, x \in T\}$, for each $x \in X_0$, $(x, X_0) \in X_1$. Hence, π_0 is surjective. Now suppose, for some $n \geq 1$, that $\pi_{n-1} : X_n \to X_{n-1}$ is surjective. We will show that π_n is surjective. Let $(x, T) \in X_n$ be arbitrary. Define $\mathcal{A} := \{(y, S) \in X_n \,|\, y \in T\}$. We will show $((x, T), \mathcal{A}) \in X_{n+1}$, which is enough, because $\pi_n((x, T), \mathcal{A}) = (x, T)$. We have shown in (6.2):

$$X_{n+1} = \{((x, T), \mathcal{A}) : (x, T) \in X_n, \mathcal{A} \subseteq X_n, (x, T) \in \mathcal{A}, T = \rho_{n-1}[\mathcal{A}]\}.$$

By assumption, $(x, T) \in X_n$, and by definition, $\mathcal{A} \subseteq X_n$. Since $(x, T) \in X_n$, we have $x \in T$, whence $(x, T) \in \mathcal{A}$, by the definition of $\mathcal{A}$. Furthermore, for all $(y, S) \in \mathcal{A}$, by definition, $y \in T$ and therefore $\rho_{n-1}[\mathcal{A}] \subseteq T$. Finally, as $T \subseteq X_{n-1}$, by the induction hypothesis, for each $y \in T$ there exists $S \subseteq X_{n-1}$ such that $(y, S) \in X_n$. Hence $T \subseteq \rho_{n-1}[\mathcal{A}]$, and we conclude $((x, T), \mathcal{A}) \in X_{n+1}$. $\qquad\square$

3.4 COLIMIT OF THE CHAIN IS THE FREE ALGEBRA

The variety $\mathcal{V}_T$ consists, by definition, of algebras from the variety $\mathcal{V}_{BA}$ of Boolean algebras, equipped with an additional operation $f = \Diamond$ which satisfies certain equations. We will now prove that the colimit of the chain defined in Subsection 3.1 is indeed the free algebra for $\mathcal{V}_T$ over the finite set of variables P. We thus obtain the finitely generated free algebra for the variety $\mathcal{V}_T$ as a colimit of a countable increasing chain of finite algebras in $\mathcal{V}$. This construction already played a crucial role in Ghilardi (1995), Bezhanishvili and Gehrke (2011) and Bezhanishvili and Kurz (2007). The proof we give here could be extended to a more general setting of varieties of modal algebras, but to ease the notation we choose to give the proof for the specific case of the variety $\mathcal{V}_T$.

Let $(\mathbb{B}_n \xrightarrow{k_n} \mathbb{B})_{n\in\mathbb{N}}$ be the colimit in the category BA of the chain diagram that we constructed in Subsection 3.1:

$$\mathbb{B}_0 \xrightarrow{i_0} \mathbb{B}_1 \xrightarrow{i_1} \cdots \qquad \mathbb{B}_n \xrightarrow{i_n} \mathbb{B}_{n+1} \xrightarrow{i_{n+1}} \cdots$$

By a theorem of Manes (1976), the colimit in BA is given by lifting the colimit in Set. Concretely, the underlying set of $\mathbb{B}$ can be described by taking the disjoint union $\bigsqcup_{n\in\mathbb{N}} B_n$, and quotienting it by the equivalence relation $\sim$, which is defined to be the smallest equivalence relation containing all pairs $\langle b_n, i_n(b_n)\rangle$, for $n \in \mathbb{N}$, $b_n \in B_n$. The Boolean algebra operations are then well-defined, and the function k_n is the inclusion of B_n into $\bigsqcup_{n\in\mathbb{N}} B_n$, followed by taking the class under $\sim$.

Now, one can show that the functions $k_{n+1} \circ \blacklozenge_n : B_n \to B$ form a cone under the diagram of which $k_n : \mathbb{B}_n \to \mathbb{B}$ is the colimit (cf. the extended version of this paper for details). By the universal property of the colimit, there exists a (unique) function, which we will denote by $\Diamond^{\mathbb{B}}$, from $\mathbb{B} \to \mathbb{B}$, such that $\Diamond^{\mathbb{B}} \circ k_n = k_{n+1} \circ \blacklozenge_n$. Concretely, the function $\Diamond^{\mathbb{B}}$ on $\mathbb{B}$ may be defined, for $b \in \mathbb{B}$, by taking some $n \in \mathbb{N}$ and $b_n \in B_n$ such that $k_n b_n = b$, and then put $\Diamond^{\mathbb{B}} b := k_{n+1} \blacklozenge_n b_n$. Using that $\blacklozenge_{n+1} \circ i_n = i_{n+1} \circ \blacklozenge_n$ for any n, it is not hard to see directly that this function $\Diamond^{\mathbb{B}}$ is well-defined.

We thus get an algebra $(\mathbb{B}, \Diamond^{\mathbb{B}})$ in the modal signature. To see that $(\mathbb{B}, \Diamond^{\mathbb{B}})$ is indeed the free $\mathcal{V}_T$-algebra over P, it remains to show the following two things:

1. $(\mathbb{B}, \Diamond^{\mathbb{B}})$ has the universal mapping property for algebras in $\mathcal{V}_T$,

2. $(\mathbb{B}, \Diamond^{\mathbb{B}})$ is in the variety $\mathcal{V}_T$.

For the proof, we refer once more to the extended version of this paper.

4 Conclusion

In this paper, we showed how to construct the Lindenbaum algebra for the modal logic T via a uniform step-by-step construction. The two main tools we used were *universal algebra*, in particular the fact that the Lindenbaum algebra for the logic T on n propositional variables is exactly the n-generated free algebra in the variety $\mathcal{V}_T$, and *Stone duality* for finite Boolean algebras, which enabled us to give a concrete set-theoretic description of the chain of algebras.

As indicated in the introduction, the actual use of uniform step-by-step constructions lies in the fact that it immediately gives normal forms: any formula in the logic of rank at most n is equivalent to the disjunction of atoms which are below it in the algebra $\mathbb{B}_n$. An interesting application of the theory in this paper would be to obtain normal forms for the modal logic T. Along the same lines, one could try implementing the uniform construction to obtain an actual algorithm for deciding logical equivalence in the modal logic T.

The idea of step-by-step constructing the free algebra originated from the case of algebras for a functor, where one can also describe the process of 'defining the next algebra in the chain' by means of a functor. For many logics, such as T and S4, the algebras for the logic are *not* algebras for a functor. However, it may still be possible to describe the process of building the chain by repeatedly applying a functor, which would then probably not be based on the category $\mathsf{BA}_{<\omega}$ itself, but on a category of certain diagrams in $\mathsf{BA}_{<\omega}$.

A more fundamental open question in this line of research is whether we can give syntactic conditions on the axioms defining a modal logic $\mathcal{L}$ which ensure that the answer to the question $(Q_{\mathcal{L}})$ from the introduction is affirmative. This will be the next direction to pursue in this research project.

Acknowledgments

This paper marked the beginning of a larger joint project with Dion Coumans (see Gool and Coumans 2012). We are very much indebted to our PhD adviser Mai Gehrke for proposing this project to us in the first place, and for her guidance along the way.

References

Bezhanishvili, N. and M. Gehrke (2011). "Finitely Generated Free Heyting Algebras via Birkhoff Duality and Coalgebra". In: *Logical Methods in Computer Science* 7.2:9, pp. 1–24.

Bezhanishvili, N. and A. Kurz (2007). "Free Modal Algebras: a Coalgebraic Perspective". In: *Algebra and Coalgebra in Computer Science*. Ed. by T. Mossakowski, U. Montanari, and M. Haveraaen. Lecture Notes in Computer Science 4624. Berlin: Springer, pp. 143–157.

Blackburn, P., M. de Rijke, and Y. Venema (2001). *Modal Logic*. Cambridge Tracts in Theoretical Computer Science 53. Cambridge University Press.

Bonsangue, M. and A. Kurz (2006). "Presenting Functors by Operations and Equations". In: *Foundations of Software Science and Computation Structures*. Ed. by L. Aceto and A. Ingólfsdóttir. Lecture Notes in Computer Science 3921. Berlin: Springer, pp. 172–186.

Burris, S. and H. P. Sankappanavar (2000). *A Course in Universal Algebra: The Millennium Edition*. Available online at http://www.math.uwaterloo.ca/~snburris/htdocs/ualg.html.

Coumans, D. and S. van Gool (2011). "Constructing the Lindenbaum Algebra for a Logic Step-by-Step Using Duality (extended version)". Available online at http://www.math.ru.nl/~vangool/coumansvangool17012011.pdf.

Ghilardi, S. (1992). "Free Heyting Algebras as Bi-Heyting Algebras". In: *Mathematical Reports of the Academy of Science*. Vol. XVI. The Royal Society of Canada, pp. 240–244.

— (1995). "An Algebraic Theory of Normal Forms". In: *Annals of Pure and Applied Logic* 71, pp. 189–245.

— (2010). "Continuity, Freeness, and Filtrations". In: *Journal of Applied Non Classical Logics* 20.3, pp. 193–217.

Gool, S. van and D. Coumans (2012). "On generalizing free algebras for a functor". In: *Journal of Logic and Computation (forthcoming)*. Preprint available online at http://www.math.ru.nl/~vangool/coumansvangool23112011.pdf.

Manes, E. G. (1976). *Algebraic Theories*. Vol. 26. Graduate Texts in Mathematics. Berlin: Springer.

Stone, M. H. (1936). "The Theory of Representation for Boolean Algebras". In: *Transactions of the American Mathematical Society* 74, pp. 37–111.

CHAPTER 7

A Logic-Based Approach to Pluralistic Ignorance

Jens Ulrik Hansen

Abstract

'Pluralistic ignorance' is a phenomenon mainly studied in social psychology. Viewed as an epistemic phenomenon, one way to define it is as a situation where *"no one believes, but everyone believes that everyone else believes"*. In this paper various versions of pluralistic ignorance are formalized using epistemic/doxastic logic (based on plausibility models). The motive is twofold. Firstly, the formalizations are used to show that the various versions of pluralistic ignorance are all consistent, thus there is nothing in the phenomenon that necessarily goes against logic. Secondly, pluralistic ignorance is on many occasions assumed to be fragile. In this paper, however, it is shown that pluralistic ignorance need not be fragile to announcements of the agents' beliefs. Hence, to dissolve pluralistic ignorance in general, something more than announcements of the subjective views of the agents is needed. Finally, suggestions to further research are outlined.

1 Introduction

Pluralistic ignorance is a term from the social and behavioral sciences going back to the work of Floyd H. Allport and Daniel Katz (Katz and Allport 1931).[1] Krech and Crutchfield (1948, pp. 388–89) define pluralistic ignorance as a situation where *"no one believes, but everyone believes that everyone else believes"*. Elaborated, pluralistic ignorance is the phenomenon where a group of people shares a false belief about

---

[1] See O'Gorman (1986) for more on the coining of the term 'pluralistic ignorance'.

the beliefs, norms, actions or thoughts of the other group members. It is a social phenomenon where people make systematic errors in judging other people's private attitudes. This makes it an important notion in understanding social life. However, pluralistic ignorance is a term used to describe many different phenomena that all share some common features. Therefore, there are many different definitions and examples of pluralistic ignorance and a few of the most common of these will be presented in Section 2.

Pluralistic ignorance has been approached by formal methods before (Centola, Willer, and Macy 2005; Hendricks 2010), but to the knowledge of the author, Hendricks (2010) is the only paper that takes a logic-based approach. Hendricks (2010) models pluralistic ignorance using formal learning theory and logic. In this paper, the tool will be classical modal logic in the form of doxastic/epistemic logic. In Section 3 we introduce a doxastic/epistemic logic based on the plausibility models presented in Baltag and Smets (2008). The reason for choosing this framework instead of, for instance, the multi-modal logic **KD45**, is that **KD45** cannot straightforwardly be combined with public announcements.[2] Since one of the aspects of pluralistic ignorance studied in this paper is the question of what it takes to dissolve the phenomenon, we need to be able to talk about the dynamics of knowledge and beliefs. Public announcements are the simplest form of actions that can affect the beliefs and knowledge of the agents and they therefore serve the purpose of this paper perfectly.

After having presented the formal framework in Section 3, it is possible in Section 4 to give a formal analysis of the different versions of pluralistic ignorance. We will give several different formalizations of pluralistic ignorance and discuss whether they are satisfiable or not. Afterwards, we will look at what it takes to dissolve pluralistic ignorance and show that, in general, something more than mere announcements of the agents' true beliefs is needed. Since the logical approach to pluralistic ignorance is still very limited, there is ample opportunity for further research and several suggestions will be discussed in Section 5. The paper ends with a concise conclusion.

2 Examples of Pluralistic Ignorance

Examples of pluralistic ignorance are plentiful in the social and behavioral sciences' literature. One example is the drinking of alcohol on (American) college campuses. Several studies have shown that many students feel much less comfortable with drinking than they believe the average college student does (Prentice and Miller 1993). In other words, the students do not believe that drinking is at all enjoyable, but they still believe that all of their fellow students believe drinking to be quite enjoyable. Another classical example is the classroom example in which, after having presented the students with difficult material, the teacher asks them whether they have any questions. Even though most students do not understand the material they may not ask any questions. All the students interpret the lack of questions from the other students as a sign that they understood the material, and to avoid being publicly displayed as the stupid one, they dare not ask questions themselves. In this case the students are ignorant with respect to some facts, but believe that the rest of the students are not ignorant about

---

[2]Public announcement of a formula φ corresponds, in the model theory of modal logic, to the operation of going to the submodel only containing worlds where φ was true. However, the class of frames underlying the logic **KD45** is not closed under taking submodels, since seriality is not preserved when going to submodels. When combined with public announcement the logic **KD45** actually turns into the logic **S5**.

the facts.

A classical made-up example is from Hans Christian Andersen's fable 'The Emperor's New Clothes' from 1837. Here, two impostors sell imaginary clothes to an emperor claiming that those who cannot see the clothes are either not fit for their office or just truly stupid. Not wanting to appear unfit for his office or truly stupid, the Emperor (as well as everyone else) pretends to be able to see the garment. No one *personally* believes the Emperor to have any clothes on. They do, however, believe that everyone else believes the Emperor to be clothed. Or alternatively, everyone is ignorant to whether the Emperor has clothes on or not, but believes that everyone else is not ignorant. Finally, a little boy cries out: "but he has nothing on at all!" and the pluralistic ignorance is dissolved.

What might be clear from these examples is that pluralistic ignorance comes in many versions. A logical analysis of pluralistic ignorance may help categorize and distinguish several of these different versions. Note that these examples were all formulated in terms of beliefs, but pluralistic ignorance is often defined in the term of norms as well. Centola, Willer, and Macy (2005) define pluralistic ignorance as "a situation where a majority of group members privately reject a norm, but assume (incorrectly) that most others accept it".

Misperceiving other people's norms or beliefs can occur without it being a case of pluralistic ignorance. Pluralistic ignorance is the case of systematic errors in norm/belief estimation of others. Thus, pluralistic ignorance is a genuine social phenomenon and not just people holding wrong beliefs about other people's norms or beliefs (O'Gorman 1986). This might be the reason why pluralistic ignorance is often portrayed as a fragile phenomenon. Just one public announcement of a private belief or norm will resolve the case of pluralistic ignorance. In 'The Emperor's New Clothes' a little boy's outcry is enough to dissolve the pluralistic ignorance. If, in the classroom example, one student dares to ask a question (and thus announces his academic ignorance) the other students will surely follow with questions of their own. In some versions of pluralistic ignorance, the mere awareness of the possibility of pluralistic ignorance is enough to suspend it. This fragility might not always be the case and, as we shall see, there is nothing in the standard definitions of pluralistic ignorance that forces it to be a fragile phenomenon.

3 Plausibility Models: A Logical Model of Belief, Knowledge, Doubt, and Ignorance

We will model knowledge and beliefs using modal logic. More specifically, we will be using the framework of Baltag and Smets (2008). This section is a review of that framework. We will work in a multi-agent setting, and thus assume a finite set of agents $\mathbb{A}$ to be given. Furthermore, we also assume a set of propositional variables PROP to be given. The models of the logic will be special kinds of Kripke models called plausibility 'models':

Definition 3.1. *A plausibility model is a tuple* $\mathcal{M} = \langle W, (\leq_a)_{a \in \mathbb{A}}, V \rangle$, *where W is a nonempty set of possible worlds/states,* $\leq_a$ *is a locally connected converse well-founded preorder on W for each* $a \in \mathbb{A}$, *and V is a valuation that to each* $p \in$ PROP *assigns a subset of W.*

A relation is a *locally connected converse well-founded preorder* on W if it is locally connected (wherever x and y are related and y and z are related, then x and z are also related), converse well-founded (every non-empty subset of W has a maximal element), and is a preorder (it is reflexive and transitive). In the following we will sometimes refer to the plausibility models simply as models.

The intuition behind plausibility models is that the possible worlds represent different ways the world might be. That $w \leq_a v$ for an agent a means that agent a thinks that the world v is at least as plausible as world w, but a cannot distinguish which of the two is the case. The relation $\leq_a$ will be used to define what agent a believes. To define what agent a knows we introduce an equivalence relation $\sim_a$ defined by:

$$w \sim_a v \quad \text{if, and only if} \quad w \leq_a v \text{ or } v \leq_a w.$$

The intuition behind $w \sim_a v$ is that for all that agent a knows, she cannot distinguish between which of the worlds w and v is the case. Given an agent a and a world w, the set $|w|_a = \{v \in W \mid v \sim_a w\}$ is the information cell at w of agent a and represents all the worlds that agent a considers possible at the world w. In other words, this set encodes the hard information of agent a at the world w.

Based on the introduced notions, we can now define knowledge and beliefs. Let K_a and B_a be modal operators for all agents $a \in \mathbb{A}$. We read $K_a\varphi$ as 'agent a knows that φ' and $B_a\varphi$ as 'agent a believes that φ'. We specify the formal language $\mathcal{L}$, which we will be working with, by the following syntax:

$$\varphi ::= p \mid \neg\varphi \mid \varphi \wedge \varphi \mid K_a\varphi \mid B_a\varphi,$$

where $p \in \mathsf{PROP}$ and $a \in \mathbb{A}$. The logical symbols $\top, \bot, \vee, \rightarrow, \leftrightarrow$ are defined in the usual way. The semantics of the logic is then defined by:

Definition 3.2. *Given a plausibility model* $\mathcal{M} = \langle W, (\leq_a)_{a \in \mathbb{A}}, V \rangle$ *and a world* $w \in W$, *we define the semantics inductively by:*

$$
\begin{array}{lll}
\mathcal{M}, w \models p & \textit{iff} & w \in V(p) \\
\mathcal{M}, w \models \neg\varphi & \textit{iff} & \text{it is not the case that } \mathcal{M}, w \models \varphi \\
\mathcal{M}, w \models \varphi \wedge \psi & \textit{iff} & \mathcal{M}, w \models \varphi \text{ and } \mathcal{M}, w \models \psi \\
\mathcal{M}, w \models K_a\varphi & \textit{iff} & \text{for all } v \in |w|_a, \ \mathcal{M}, v \models \varphi \\
\mathcal{M}, w \models B_a\varphi & \textit{iff} & \text{for all } v \in max_{\leq_a}(|w|_a), \ \mathcal{M}, v \models \varphi,
\end{array}
$$

where $max_{\leq_a}(S)$ *is the set of maximal elements of S with respect to the relation $\leq_a$. We say that a formula φ is satisfiable if there is a model $\mathcal{M}$ and a world w in $\mathcal{M}$ such that* $\mathcal{M}, w \models \varphi$. *A formula φ is valid if for all models $\mathcal{M}$ and all worlds w in $\mathcal{M}$,* $\mathcal{M}, w \models \varphi$.

Note that the semantics make $K_a\varphi \rightarrow B_a\varphi$ valid. In the framework of Baltag and Smets (2008) other notions of beliefs are also introduced. The first is *conditional belief*: $B_a^\varphi\psi$ expresses that agent a believes that ψ was the case, if she learned that φ was the case. The semantics of this modality is:

$$\mathcal{M}, w \models B_a^\varphi\psi \quad \text{iff} \quad \text{for all } v \in max_{\leq_a}(|w|_a \cap [\![\varphi]\!]_{\mathcal{M}}), \ \mathcal{M}, v \models \psi,$$

where $[\![\varphi]\!]_{\mathcal{M}}$ is the set of worlds in $\mathcal{M}$ where φ is true. Another notion of belief is *safe belief* for which we use $\Box_a$. The semantics of this modality is:

$$\mathcal{M}, w \models \Box_a\varphi \quad \text{iff} \quad \text{for all } v \in W, \text{ if } w \leq_a v, \text{ then } \mathcal{M}, v \models \varphi.$$

Note that this is the usual modality defined from the relation $\leq_a$. Since $\leq_a$ is reflexive, $\Box_a\varphi \to \varphi$ is valid. Hence, safe belief is a very strong notion of belief (or a weak notion of knowledge). Because a central aspect of pluralistic ignorance is people holding *wrong* beliefs, safe belief is not a suitable notion. Yet another notion of belief, that also implies truth, is *weakly safe belief* $\Box_a^{\text{weak}}$ given by the following semantics:

$$\mathcal{M}, w \models \Box_a^{\text{weak}}\varphi \quad \text{iff} \quad \mathcal{M}, w \models \varphi \text{ and for all } v \in W,$$
$$\text{if } w <_a v, \text{ then } \mathcal{M}, v \models \varphi,$$

where $<_a$ is defined by: $w <_a v$ if and only if $w \leq_a v$ and $v \not\leq_a w$. Finally, Baltag and Smets (2008) define *strong belief* Sb_a by

$$Sb_a\varphi \text{ iff } B_a\varphi \wedge K_a(\varphi \to \Box_a\varphi).$$

In addition to the several notions of belief, Baltag and Smets (2008) also discuss several ways of updating knowledge and beliefs when new information comes about. These are *update*, *radical upgrade*, and *conservative upgrade* and can be distinguished by the trust that is put in the source of the new information. If the source is known to be infallible, it should be an update. If the source is highly reliable, it should be a radical upgrade, and if the source is just barely trusted, it should be a conservative upgrade. In this paper we are interested in what it takes to dissolve pluralistic ignorance and since update is the 'strongest' way of updating knowledge and beliefs, we will focus on this. We will also refer to this way of updating as *public announcement*.

We introduce operators $[!\varphi]$, and add to the syntax the clause that for all formulas φ and ψ, $[!\varphi]\psi$ is also a formula. $[!\varphi]\psi$ is read as 'after an announcement of φ, ψ is true'. Semantically, a public announcement of φ will result in a new plausibility model where all the $\neg\varphi$-worlds have been removed, and the truth of ψ is then checked in this new model. These intuitions are made formal in the following definition:

Definition 3.3. *Given a plausibility model* $\mathcal{M} = \langle W, (\leq_a)_{a\in\mathbb{A}}, V \rangle$ *and a formula* φ, *we define a new model* $\mathcal{M}_{!\varphi} = \langle W', (\leq_a')_{a\in\mathbb{A}}, V' \rangle$ *by,*

$$
\begin{aligned}
W' &= \{w \in W \mid \mathcal{M}, w \models \varphi\}, \\
\leq_a' &= \leq_a \cap (W' \times W'), \\
V'(p) &= V(p) \cap W'.
\end{aligned}
$$

The semantics of the public announcement formulas are then given by:

$$\mathcal{M}, w \models [!\varphi]\psi \quad \textit{iff} \quad \textit{if } \mathcal{M}, w \models \varphi \textit{ then } \mathcal{M}_{!\varphi}, w \models \psi.$$

Finally, we add to the framework of Baltag and Smets (2008) the two notions of ignorance and doubt. These are notions rarely discussed in the literature on epistemic/doxastic logic. However, ignorance is discussed in van der Hoek and Lomuscio (2004). On the syntactic level we add two new operators I_a and D_a for each agent $a \in \mathbb{A}$. The formula $I_a\varphi$ is read as 'agent a is ignorant about φ' and $D_a\varphi$ is read as 'agent a doubts whether φ'. The semantics of these operators are defined from the semantics of the knowledge operator and the belief operator:

Definition 3.4. *The operators* I_a *and* D_a *are defined be the following equivalences:*
$$
\begin{aligned}
I_a\varphi &:= \neg K_a\varphi \wedge \neg K_a\neg\varphi \\
D_a\varphi &:= \neg B_a\varphi \wedge \neg B_a\neg\varphi.
\end{aligned}
$$

Note that since $K_a\varphi \to B_a\varphi$ is valid, $D_a\varphi \to I_a\varphi$ is also valid.

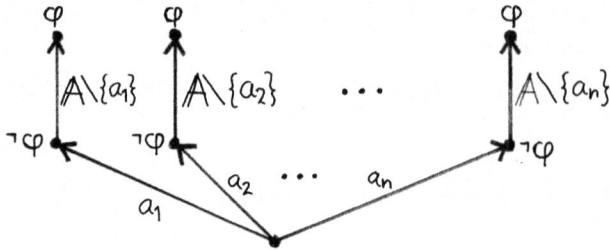

Figure 7.1: A plausibility model where (7.1) is satisfiable at the root.

4 Modeling Pluralistic Ignorance

Based on the logic introduced in the previous section, we will now formalize different versions of pluralistic ignorance that are all consistent. Then, we will discuss whether these formalizations make pluralistic ignorance into a fragile phenomenon.

4.1 FORMALIZATIONS AND CONSISTENCY OF PLURALISTIC IGNORANCE

As discussed in Section 2, there are many ways of defining pluralistic ignorance and in this section we attempt to formalize a few of these. We will also discuss whether these formalizations lead to consistent concepts in the sense that the formalizations are by satisfiable formulas.

Firstly, we assume that pluralistic ignorance is a situation where no agent believes φ, but every agent believes that everyone else believes φ. This can easily be formalized as:

$$\bigwedge_{a \in \mathbb{A}} \left(\neg B_a \varphi \wedge \bigwedge_{b \in \mathbb{A} \setminus \{a\}} B_a B_b \varphi \right). \tag{7.1}$$

For boolean formulas φ,[3] (7.1) is satisfiable since a plausibility model can easily be constructed such that it contains a possible world that satisfies it. Such a model is given in Figure 7.1, where we assume that the set of agents is $\mathbb{A} = \{a_1, a_2, ..., a_n\}$. In the following, when drawing models like this one, an arrow from a state w to a state v labeled by a_i will represent that $w <_{a_i} v$ holds in the model. An arrow from w to v labeled by a set $B \subseteq \mathbb{A}$ represent that $w <_b v$ for all $b \in B$. The full plausibility relations of the model will be the reflexive transitive closures of the relations drawn in the pictures. When a formula φ appears next to a state it means that φ is true at that state.

There are also formulas φ for which (7.1) is unsatisfiable, take for instance φ to be $B_b \psi$ or $\neg B_b \psi$ for any agent $b \in \mathbb{A}$ and any formula ψ. It cannot be the case that agent a does not believe that agent b believes that ψ, but at the same time a believes that b believes that b believes that ψ, i.e. (7.1) is unsatisfiable when φ is $B_b \psi$ or $\neg B_b \psi$ because $\neg B_a B_b \psi \wedge B_a B_b B_b \psi$ and $\neg B_a \neg B_b \psi \wedge B_a B_b \neg B_b \psi$ are unsatisfiable. In the following, when discussing pluralistic ignorance as defined by (7.1) we will therefore assume that φ is a boolean formula.

If belief is replaced by strong belief, such that (7.1) becomes

---

[3]A formula is boolean if it constructed solely from propositional variables and the logical connectives $\neg$, $\wedge$, $\vee$, $\rightarrow$, and $\leftrightarrow$.

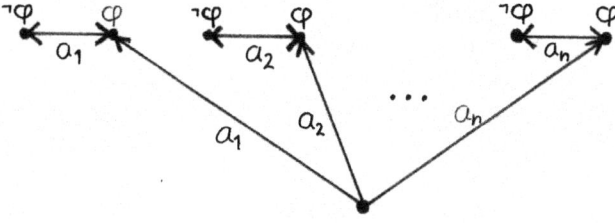

Figure 7.2: A plausibility model where (7.3) and (7.4) are satisfiable at the root.

$$\bigwedge_{a\in A}\left(\neg Sb_a\varphi \wedge \bigwedge_{b\in A\backslash\{a\}}Sb_aSb_b\varphi\right),\tag{7.2}$$

pluralistic ignorance remains satisfiable for boolean formulas, which is testified by Figure 7.1 again. Furthermore, (7.2) is also not satisfiable if φ is of the form $Sb_b\psi$ or $\neg Sb_b\psi$ for a $b \in A$. However, if we use safe belief and weak safe belief instead of belief in (7.1), pluralistic ignorance becomes unsatisfiable. This is obvious since both safe belief and weak safe belief imply truth.

In the classroom example of Section 2, a better definition of pluralistic ignorance may be obtained using the ignorance operator. This leads to the following definition of pluralistic ignorance:

$$\bigwedge_{a\in A}\left(I_a\varphi \wedge \bigwedge_{b\in A\backslash\{a\}}B_a\neg I_b\varphi\right).\tag{7.3}$$

This formula expresses a case where all the agents are ignorant about φ, but believe that all the other agents are not ignorant about φ. Instead of ignorance, doubt could be used as well, providing yet another definition of pluralistic ignorance:

$$\bigwedge_{a\in A}\left(D_a\varphi \wedge \bigwedge_{b\in A\backslash\{a\}}B_a\neg D_b\varphi\right).\tag{7.4}$$

Note that, since $D_a\varphi \rightarrow I_a\varphi$, (7.4) implies (7.3).

The two definitions of pluralistic ignorance (7.3) and (7.4) are also easily seen to be satisfiable for boolean formulas φ. This is made apparent by Figure 7.2. Now, however, formulas of the form $B_b\varphi$, for $b \in A$, can also be subject to pluralistic ignorance. It is possible that agent a can doubt whether agent b believes φ and at the same time believe that agent b does not doubt whether he himself /agent b believes φ.

In (7.3) and (7.4) we can also replace the belief operator by the strong belief operator and obtain the following versions of pluralistic ignorance:

$$\bigwedge_{a\in A}\left(I_a\varphi \wedge \bigwedge_{b\in A\backslash\{a\}}Sb_a\neg I_b\varphi\right),\tag{7.5}$$

$$\bigwedge_{a\in A}\left(D_a\varphi \wedge \bigwedge_{b\in A\backslash\{a\}}Sb_a\neg D_b\varphi\right).\tag{7.6}$$

These new definitions of pluralistic ignorance are consistent as they are satisfiable at the root of the model in Figure 7.2. We still cannot obtain versions of (7.3) and (7.4) with safe belief and weak safe belief for the same reason as before.

It seems obvious that we can formalize even further versions of pluralistic ignorance within this framework. Thus, using the logic introduced in Section 3, we can characterize and distinguish many different versions of pluralistic ignorance. Furthermore, all the definitions (7.1)-(7.6) were satisfiable, which seems to entail that the concept of pluralistic ignorance is not inconsistent.

4.2 THE FRAGILITY OF PLURALISTIC IGNORANCE

After having formalized different versions of pluralistic ignorance, we can ask whether any of the definitions entail that pluralistic ignorance is a fragile phenomenon. However, first of all we need to spell out what we mean by a fragile phenomenon. The question of whether pluralistic ignorance is fragile or not reduces to the question of what it takes to dissolve it. We will regard pluralistic ignorance as dissolved only when none of the agents have wrong beliefs about the other agents' beliefs anymore. The way agents can change their beliefs will in this section be modeled by the $[!\varphi]$ operators of Section 3.

For the time being, we fix pluralistic ignorance to be defined as (7.1). According to several descriptions of pluralistic ignorance, it should be dissolved if just one agent announces his true beliefs. If the formula $!\neg B_b\varphi$ is announced, it naturally follows that $\bigwedge_{a\in \mathbb{A}} B_a\neg B_b\varphi$. However, this does not dissolve the pluralistic ignorance since all agents might keep their wrong beliefs about any other agent than b. In other words, a model satisfying (7.1) can be constructed such that after the announcement of $!\neg B_b\varphi$ it still holds that $\bigwedge_{a\in \mathbb{A}} \left(\bigwedge_{c\in \mathbb{A}\setminus\{a,b\}} B_a B_c\varphi \right)$.

It turns out that there is nothing in the definition (7.1) that prevents the wrong beliefs of the agents from being quite robust. Even if everybody except an agent c announces that they do not believe φ, all the agents might still believe that c believes φ. Using a formula of $\mathcal{L}$ we can define a notion of robustness in the following way: *agent a robustly believes that the group of agents $B \subseteq \mathbb{A}\setminus\{a\}$ believes φ* if

$$\bigwedge_{C\subseteq B} \left([!\neg B_c\varphi]_{c\in C}(\bigwedge_{b\in B\setminus C} B_a B_b\varphi) \right), \tag{7.7}$$

where $[!\neg B_c\varphi]_{c\in C}$ is an abbreviation for $[!\neg B_{c_1}\varphi][!\neg B_{c_2}\varphi]...[!\neg B_{c_k}\varphi]$, when $C = \{c_1, c_2, ..., c_k\}$.[4] An example of a model where agent 1 believes $\neg\varphi$ and robustly believes that the agents $\{2, 3, 4, 5\}$ believe φ is shown in Figure 7.3.

Another way of looking at the formula (7.7) is that it describes a situation where agent a believes that all the other agents' beliefs about φ are independent; maybe they all believe φ for different reasons. Thus, learning about some agents' beliefs about φ tells a nothing about what the other agents believe about φ.

---

[4] An alternative to (7.7) is

$$\bigwedge_{C\subseteq B} \left([!\bigwedge_{c\in C}\neg B_c\varphi](\bigwedge_{b\in B\setminus C} B_a B_b\varphi) \right),$$

however, the two are not equivalent. We will not go into a discussion of which definition is preferable.

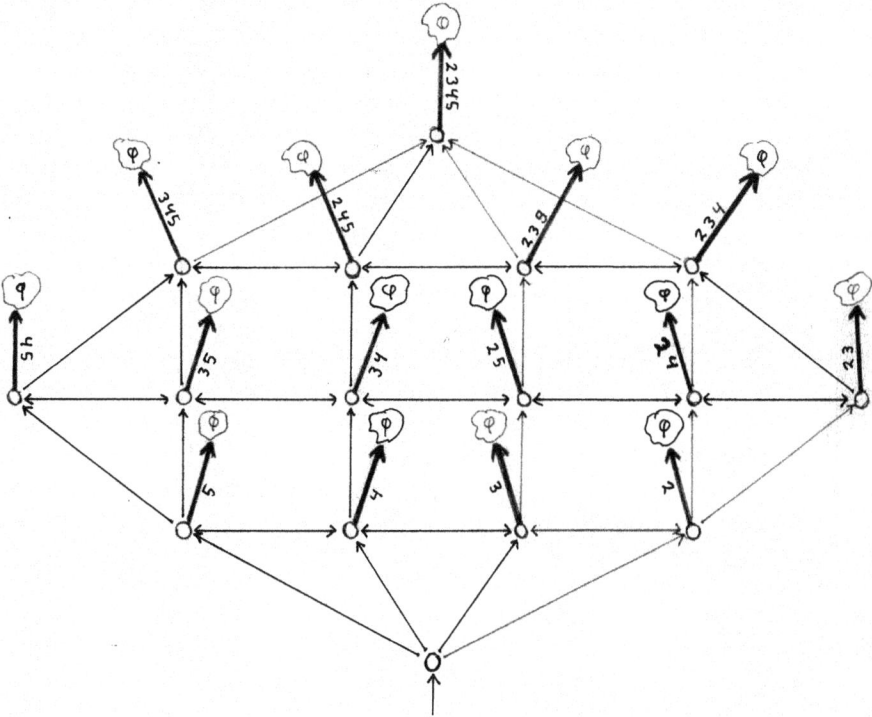

Figure 7.3: A robust model where agent 1 believes $\neg\varphi$ and has a strong robust belief in that the agents 2, 3, 4, and 5 believe φ. The worlds marked with "○" are worlds where $\neg\varphi$ is true and the "clouds" marked with φ are collections of worlds where φ is true all over. The arrows not marked with numbers represent the plausibility relation for agent 1 only.

With robustness defined by (7.7), pluralistic ignorance is consistent with all the agents having wrong robust beliefs about the other agents' beliefs. Taking disjoint copies of the model in Figure 7.3 for each agent and joining the roots shows that:

Proposition 4.1. *Pluralistic ignorance in form of (7.1) is consistent with all the agents $a \in \mathbb{A}$ robustly believing that the group of agents $\mathbb{A}\backslash\{a\}$ believes φ.*

Another way of interpreting this result is that announcements of the true beliefs of some of the involved agents are not enough to dissolve pluralistic ignorance. Either all the agents need to announce their true beliefs or new information has to come from an outside trusted source. Thus, announcements of the forms $!B_a\varphi$ or $!\neg B_a\varphi$ are not guaranteed to dissolve pluralistic ignorance. However, a public announcement of $!\neg\varphi$ in the model of figure 7.3 will remove the pluralistic ignorance. But an announcement of the form $!\neg\varphi$ (or $!\varphi$) is precisely an announcement from an trusted outsider. An agent a in $\mathbb{A}$ can only announce formulas of the form $!B_a\psi$ or $!\neg B_a\psi$.

What turns pluralistic ignorance into a fragile phenomenon in most cases, is the fact that the agents consider the other agents' beliefs not to be independent as is the case if (7.7) is satisfied. In other words, pluralistic ignorance in the fragile form occurs mainly when the beliefs of the involved agents are correlated. This fits well with the view that pluralistic ignorance is a genuine social phenomenon as claimed by O'Gorman (1986).

Proposition 4.1 only regards pluralistic ignorance as defined by (7.1). However, for the definitions (7.3) and (7.4) similar results hold. Neither of the definitions (7.3) and (7.4) entail that pluralistic ignorance is fragile to public announcements of doubts ($[!D_b\varphi]$) or ignorance ($[!I_b\varphi]$). We can construct a new model, similar to the one in Figure 7.3, in which an agent a doubts whether φ but has a strong robust belief in that all the agents in $\mathbb{A}\backslash\{a\}$ do not doubt whether φ (and the same goes for ignorance). This new model is shown in Figure 7.4.

When it comes to the definitions of pluralistic ignorance based on strong beliefs (7.2), (7.5), and (7.6), something interesting happens. In the model of Figure 7.3 agent 1 does not have a strong belief that the other agents have strong beliefs in φ. For instance, there is a state where $B_1Sb_5\varphi$ and $Sb_5\varphi$ are true, but $\square_1Sb_5\varphi$ is not true. The same issue occurs in the model of figure 7.4. It is still unknown whether robust models can be constructed such that they satisfy the strong belief versions of pluralistic ignorance as defined by (7.2), (7.5), and (7.6). Thus, it is left for further research whether there are strong belief versions of pluralistic ignorance that are not fragile. There are several other questions for further research, which we will turn to now.

5 Further Research on Logic and Pluralistic Ignorance

We have given several consistent formalizations of pluralistic ignorance, but there still seems to be more possible variations to explore. Furthermore, we have been working within one specific framework, and the question remains of whether there are other natural frameworks in which all formalizations of pluralistic ignorance become inconsistent. This would be highly unexpected though. Another question regarding formalizations of pluralistic ignorance in different frameworks is whether it changes the fragility of the phenomenon. This is still an open question.

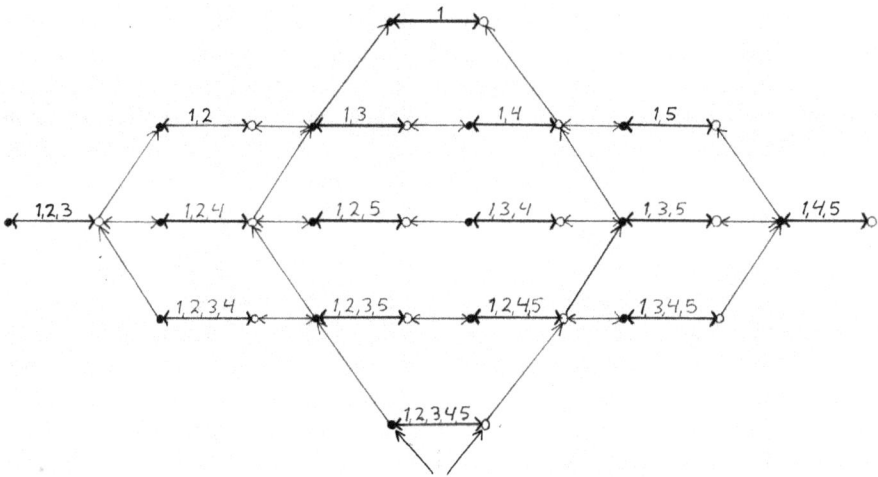

Figure 7.4: A robust model where agent 1 doubts whether φ, has a strong robust belief in that the agents 2, 3, 4, and 5 do not doubt whether φ. The worlds marked with "∘" are worlds where $\neg\varphi$ is true and the worlds marked with "•" are worlds where φ is true. The arrows with no numbers on are arrows for agent 1. Remeber that the full plausibility relations of the model are the reflexive transitive closures of the arrows in the pictures

Even though pluralistic ignorance need not be fragile, neither as a 'real life' phenomenon nor according to the formalizations given in this paper, it seems that the really interesting cases occur when pluralistic ignorance is, in fact, fragile. Whether pluralistic ignorance is fragile appears to be closely related to it being a genuine social phenomenon; the dependence between agents' beliefs is what makes pluralistic ignorance fragile. Thus, the real interesting question for further research is how agents' beliefs are interdependent in the case of pluralistic ignorance and how best to model this in logic. In answering this question, a shift in focus from what it takes to dissolve pluralistic ignorance to what it takes for pluralistic ignorance to arise, seems natural.

5.1 INFORMATIONAL CASCADES: HOW PLURALISTIC IGNORANCE COMES ABOUT AND HOW IT VANISHES

An agent's beliefs may depend on other agents' beliefs in many ways; one way is through testimony of facts by other agents in which the agent trusts. Modeling trust and testimony is for instance done in Holliday (2010). Another way in which agents' beliefs may depend on each other could be through a common information source (Bikhchandani, Hirshleifer, and Welch 1998). Yet another way is through informational cascades.

Informational cascades are phenomena widely discussed in the social sciences Bikhchandani, Hirshleifer, and Welch 1998; Lohmann 1994 and the term was introduced by Bikhchandani, Hirshleifer, and Welch 1992. Assume that some agents are supposed to act one at a time in a given order and that their actions depend on a private information source as well as the information obtained by observing the actions of the agents already having acted. When actions are performed sequentially and agents start to ignore their private information and instead base their actions merely on information obtained from the actions of the previous agents, an *informational cascade* has occurred. If the actions of the first people in the cascade oppose to their private beliefs and the remaining people join in with the same actions (also oppose to their private beliefs) the result might be a case of pluralistic ignorance. However, informational cascades are also fragile Bikhchandani, Hirshleifer, and Welch 1992 and opposite cascades may occur, thus eliminating pluralistic ignorance again.

These kinds of informational cascades, which have been shown to occur in numerous places, may very well be the cause of pluralistic ignorance. Hence, a logical framework that can model informational cascades might also be suited to model pluralistic ignorance. To the knowledge of the author, the only paper on logic-based models of informational cascades is Holliday (2010), but it may very well be possible to model pluralistic ignorance in that framework. However, further work on the logics of informational cascades is still to come.

5.2 PRIVATE VERSUS PUBLIC BELIEFS – THE NEED FOR NEW NOTIONS OF GROUP BELIEFS

The concept of pluralistic ignorance, regardless of which version one adopts, seems to hint at the need for new notions of common knowledge/beliefs. Pluralistic ignorance can be viewed as a social phenomenon where everybody holds a private belief in φ, but publicly display a belief in $\neg\varphi$ and thus contribute to a 'common belief' ('public belief' might be a better word) in $\neg\varphi$. Due to the usual definition of common belief (everybody believes φ and everybody believes that everybody believes φ and ...), a common belief in $\neg\varphi$ leads to private belief in $\neg\varphi$ for all agents in the group, but this is exactly the thing that fails in social epistemic scenarios involving pluralistic ignorance. Hence, a new notion of common group belief seems to be needed. In general, there are various ways in which group beliefs can be related to the beliefs of individuals of the group. Thus, a logic that distinguishes between private and public beliefs or contains new notions of common beliefs may help model pluralistic ignorance more adequately. Once again, this is left for further research.

5.3 HOW AGENTS ACT

The way agents act in cases of pluralistic ignorance also seems to play an important role. The reason why most students believe that other students are comfortable with drinking might be that they observe the other students drinking heavily. In the classroom example students are also obtaining their wrong beliefs based on the observation of others. Furthermore, focusing on actions might also tell us something about how pluralistic ignorance evolves in the first place.

Therefore, a logic combining beliefs and actions might be the natural tool for modeling pluralistic ignorance. There exist several logics that combine beliefs/knowledge and actions, but which one to choose and the actual modeling is left for further research to decide.

6 Conclusion

Firstly, we have seen that there are many ways of defining pluralistic ignorance, all of which by satisfiable formulas. Therefore, pluralistic ignorance is (seemingly) not a phenomenon that goes against logic. In other words, wrong logical reasoning is not necessarily involved in pluralistic ignorance.

Secondly, the standard definitions of pluralistic ignorance, for instance as a situation where no one believes, but everyone believes that everyone else believes, do not entail that the phenomenon is fragile. Public announcements of the true beliefs of some of the involved agents are not enough to dissolve pluralistic ignorance. Either all the agents need to announce their true beliefs or new information has to come from an outside, trusted source. However, pluralistic ignorance often seems to occur in cases where the agents' beliefs are correlated and in such cases pluralistic ignorance might be increasingly more fragile.

The paper has hinted at a first logic approach to pluralistic ignorance. Some features and problems have been singled out, but the main aim of the paper was to pave the way for further research into logical modeling of social phenomena such as pluralistic ignorance.

Acknowledgments

The author would like to thank the participants of the workshop PhDs in Logic III (Brussels, February 17-18th, 2011) and the participants of the Copenhagen-Lund workshop in Social Epistemology (Lund, February 25th, 2011). Furthermore, the author would like to thank Jens Christian Bjerring, Nikolaj Jang Lee Linding Pedersen, Carlo Proietti, Vincent F. Hendricks, Eric Pacuit, and Olivier Roy for useful discussions and comments.

References

Baltag, A. and S. Smets (2008). "A Qualitative Theory of Dynamic Interactive Belief Revision". In: *Logic and the Foundations of Game and Decision Theory (LOFT7)*. Ed. by G. Bonanno, W. van der Hoek, and M. Wooldridge. Texts in Logic and Games 3. Amsterdam: Amsterdam University Press, pp. 13–60.

Bikhchandani, S., D. Hirshleifer, and I. Welch (1992). "A Theory of Fads, Fashion, Custom, and Cultural Change as Informational Cascades". In: *Journal of Political Economy* 100, pp. 992–1026.

— (1998). "Learning from the Behavior of Others: Conformity, Fads, and Informational Cascades". In: *Journal of Economic Perspectives* 12, pp. 151–170.

Centola, D., R. Willer, and M. Macy (2005). "The Emperor's Dilemma. A Computational Model of Self-Enforcing Norms". In: *American Journal of Sociology* 110, pp. 1009–1040.

Hendricks, V. F. (2010). "Knowledge Transmissibility and Pluralistic Ignorance: A First Stab". In: *Metaphilosophy* 41, pp. 279–291.

Holliday, W. H. (2010). "Trust and the Dynamics of Testimony". In: *Logic and Interactive Rationality – Seminar's Yearbook 2009*. Ed. by D. Grossi, L. Kurzen, and F. Velázquez-Quesada. Institute for Logic, Language and Computation, Universiteit van Amsterdam, pp. 147–178.

Katz, D. and F. H. Allport (1931). *Student Attitudes*. Syracuse, NY: The Craftsman Press.

Krech, D. and R. S. Crutchfield (1948). *Theories and Problems of Social Psychology*. New York, NY: McGraw-Hill.

Lohmann, S. (1994). "The Dynamics of Informational Cascades: The Monday Demonstrations in Leipzig, East Germany, 1989-91". In: *World Politics* 47.10, pp. 42–101.

O'Gorman, H. J. (1986). "The Discovery of Pluralistic Ignorance: An Ironic Lesson". In: *Journal of the History of the Behavioral Sciences* 22, pp. 333–347.

Prentice, D. A. and D. T. Miller (1993). "Pluralistic Ignorance and Alcohol Use on Campus: Some Consequences of Misperceiving the Social Norm". In: *Journal of Personality and Social Psychology* 64, pp. 243–256.

van der Hoek, W. and A. Lomuscio (2004). "A Logic For Ignorance". In: *Electronic Notes in Theoretical Computer Science* 85, pp. 117–133.

CHAPTER 8

Groups with Unbounded Potential Automorphism Tower Heights

Philipp Lücke

Abstract

We show that it is consistent with the axioms of ZFC that there exists an infinite centreless group G with the property that for every ordinal α there is a notion of forcing $\mathbb{P}$ that preserves cardinalities and cofinalities and forces the automorphism tower of G to be taller than α.

1 Introduction

In this note, we construct a model of set theory that contains a centreless group with *unbounded potential automorphism tower heights*. Our set-theoretic notation is standard and follows Kunen (1980).

We start by giving a brief introduction to the so-called *automorphism tower problem*.[1] Let G be a group with trivial centre. For each $g \in G$, the map

$$\iota_g : G \longrightarrow G : h \mapsto g \circ h \circ g^{-1}$$

is an automorphism of G and is called the *inner automorphism corresponding to g*. Clearly, $\iota_g = \mathrm{id}_G$ if and only if $g = \mathbb{1}_G$. The map

$$\iota_G : G \longrightarrow \mathrm{Aut}(G) : g \mapsto \iota_g$$

---

[1] An extensive account of all aspects of the automorphism tower problem can be found in Simon Thomas' forthcoming monograph *The Automorphism Tower Problem*.

is an embedding of groups that maps G onto the subgroup $\mathrm{Inn}(G)$ of all inner automorphisms of G. An easy computation shows that $\pi \circ \iota_g \circ \pi^{-1} = \iota_{\pi(g)}$ holds for all $g \in G$ and $\pi \in \mathrm{Aut}(G)$. This shows that $\mathrm{Inn}(G)$ is a normal subgroup of $\mathrm{Aut}(G)$ and $\mathrm{Aut}(G)$ is also a group with trivial centre. By iterating this process, we inductively construct the automorphism tower of G.

Definition 1.1. *A sequence $\langle G_\alpha \mid \alpha \in \mathrm{On} \rangle$ of groups is the automorphism tower of a centreless group G if the following statements hold.*

1. $G_0 = G$.

2. *For all $\alpha \in \mathrm{On}$, G_α is a normal subgroup of $G_{\alpha+1}$ and the induced homomorphism*

$$\varphi_\alpha \colon G_{\alpha+1} \longrightarrow \mathrm{Aut}(G_\alpha) \colon g \mapsto \iota_g \upharpoonright G_\alpha$$

is an isomorphism.

3. *For all $\alpha \in \mathrm{Lim}$, $G_\alpha = \bigcup \{G_\beta \mid \beta < \alpha\}$.*

In this definition, we replaced $\mathrm{Aut}(G_\alpha)$ by an isomorphic copy $G_{\alpha+1}$ that contains G_α as a normal subgroup. This allows us to take unions at limit stages. Without this isomorphic correction, we would have to take direct limits at limit stages. By induction, we can construct such a tower for each centreless group and it is easy to show that each group G_α in this tower is uniquely determined up to an isomorphism which is the identity on G. We can therefore speak of *the* α-th group G_α in the automorphism tower of a centreless group G.

It is natural to ask whether the automorphism tower of a centreless group eventually *terminates*, in the sense that there is an $\alpha \in \mathrm{On}$ with $G_\alpha = G_{\alpha+1}$ and therefore $G_\alpha = G_\beta$ for all $\beta \geq \alpha$. A classical result of Helmut Wielandt (see Wielandt 1939, page 212) says that the automorphism tower of a finite centreless group terminates after finitely many steps. In Thomas (1985) and Thomas (1998), Simon Thomas uses Fodor's Lemma to show that every infinite centreless group has a terminating automorphism tower by proving the following result.

Theorem 1.2 (Thomas 1998, Theorem 1.3). *If G is an infinite centreless group of cardinality κ, then there is an $\alpha < (2^\kappa)^+$ with $G_\alpha = G_{\alpha+1}$.*

This result allows us to make the following definitions.

Definition 1.3. *Given a centreless group G, we let $\tau(G)$ denote the least ordinal α satisfying $G_\alpha = G_{\alpha+1}$ and call this ordinal the height of the automorphism tower of G. For every infinite cardinal κ, we define*

$$\tau_\kappa = \mathrm{lub}\{\tau(G) \mid G \text{ is a centreless group of cardinality } \kappa\}.$$

There are only 2^κ-many centreless groups of infinite cardinality κ and this shows that Simon Thomas' result implies $\tau_\kappa < (2^\kappa)^+$ for all infinite cardinals κ. The following result of Winfried Just, Saharon Shelah and Simon Thomas shows that $(2^\kappa)^+$ is the best *upper cardinal bound* for τ_κ provable in ZFC for uncountable regular κ.

Theorem 1.4 (Just, Shelah, and Thomas 1999, Theorem 1.4). *Assume* (GCH). *Let κ be a regular uncountable cardinal, ν be a cardinal with $\kappa < \mathrm{cof}(\nu)$ and α be an ordinal with $\alpha < \nu^+$. Then there exists a $<\kappa$-closed partial order $\mathbb{P}$ that satisfies the κ^+-chain condition and the following statements are true in every $\mathbb{P}$-generic extension of* V.

1. $2^\kappa = \nu$.

2. *There exists a centreless group G of cardinality κ such that $\tau(G) = \alpha$.*

Note that the following problem is still open.

Problem 1.5. *Find a model $\langle M, \in_M \rangle$ of ZFC and an infinite cardinal κ in M such that it is possible to compute[2] the exact value of τ_κ in M.*

One of the reasons why it is so difficult to compute the value of τ_κ is that although the definition of automorphism towers is purely algebraic, there can be groups whose automorphism tower heights depend on the model of set theory in which they are computed. Therefore, one always has to take into account the set-theoretic background in which the computation of τ_κ takes place. This shows that the automorphism tower construction contains a *set-theoretic essence* (this formulation is due to Joel David Hamkins, see Hamkins 2002). We give a short overview on results concerning the existence of such groups and continue by introducing a new class of groups whose automorphism towers are highly malleable by forcing.

In Thomas (1998), Simon Thomas constructs a centreless group G with $\tau(G) = 0$ and a partial order $\mathbb{P}$ that satisfies the countable chain condition and $\mathbb{1}_\mathbb{P} \Vdash_\mathbb{P} "\tau(\check{G}) \geq 1"$. In the other direction, he also constructs a centreless group H with $\tau(H) = 2$ and $\mathbb{1}_\mathbb{Q} \Vdash_\mathbb{Q} "\tau(\check{H}) = 1"$ for every notion of forcing $\mathbb{Q}$ that adds a new real.

Let G be an infinite centreless group, $\mathbb{P}$ be a partial order and F be $\mathbb{P}$-generic over V. By the results mentioned above, the height of the automorphism tower of G computed in V, $\tau(G)^V$, can be higher or smaller than the height computed in V[F], $\tau(G)^{V[F]}$. It is natural to ask whether the value of $\tau(G)^V$ places any constraints on the value of $\tau(G)^{V[F]}$, and vice versa. Obviously, $\tau(G)^{V[F]} = 0$ implies $\tau(G)^V = 0$. The following result by Joel David Hamkins and Simon Thomas suggests that this is the only implication provable in ZFC that holds for all centreless groups in the above situation.

Theorem 1.6 (Hamkins and Thomas 2000, Theorem 1.4). *It is consistent with the axioms of ZFC that for every infinite cardinal κ and every ordinal $\alpha < \kappa$, there exists a centreless group G with the following properties.*

1. $\tau(G) = \alpha$.

2. *Given $0 < \beta < \kappa$, there exists a partial order $\mathbb{P}$, which preserves cofinalities and cardinalities, such that $\mathbb{1}_\mathbb{P} \Vdash_\mathbb{P} "\tau(\check{G}) = \check{\beta}"$.*

In Fuchs and Hamkins (2008), Gunter Fuchs and Joel David Hamkins showed that Gödel's constructible universe L is a model of the above statement. Using techniques developed in Hamkins and Thomas (2000) and Fuchs and Hamkins (2008), Gunter

---

[2] By '*compute*', we mean *set-theoretic* characterizations of τ_κ. Examples of such characterizations are $\langle M, \in_M \rangle \models "\tau_\kappa = \kappa^+"$ or $\langle M, \in_M \rangle \models "\tau_\kappa = 2^\kappa"$.

Fuchs and the author extended the above results by constructing ZFC-models containing groups whose automorphism tower height can changed again and again by passing to another model of set theory.

Theorem 1.7 (Fuchs and Lücke 2011). *It is consistent with the axioms of ZFC that for every infinite cardinal κ there is a centreless group G with $\tau(G) = 0$ and the property that for every function $s \colon \kappa \longrightarrow (\kappa \setminus \{0\})$ there is a sequence $\langle \mathbb{P}_\alpha \mid 0 < \alpha < \kappa \rangle$ of partial orders such that the following statements hold.*

1. *For all $0 < \alpha < \kappa$, $\mathbb{P}_\alpha$ preserves cardinalities and cofinalities.*

2. *For all $0 < \alpha < \beta < \kappa$, there is a partial order $\mathbb{Q}$ with $\mathbb{P}_\beta = \mathbb{P}_\alpha \times \mathbb{Q}$.*

3. *For all $\alpha < \kappa$, we have $\mathbb{1}_{\mathbb{P}_{\alpha+1}} \Vdash_{\mathbb{P}_{\alpha+1}}$ " $\tau(\check{G}) = \check{s}(\check{\alpha})$ ".*

4. *If $0 < \alpha < \kappa$ is a limit ordinal, then $\mathbb{1}_{\mathbb{P}_\alpha} \Vdash_{\mathbb{P}_\alpha}$ " $\tau(\check{G}) = 1$ ".*

Again, L is a model of this statement. In another direction, Fuchs and Lücke (2011) also shows how to construct a model of ZFC that contains an infinite centreless group whose automorphism tower height can be changed again and again by passing to smaller and smaller inner models.

Next, we introduce another class of groups whose automorphism tower height can be changed drastically by passing to forcing extensions of the ground model. We state the main result of this note.

Theorem 1.8. *It is consistent with the axioms of ZFC that there is a centreless group G of cardinality $\aleph_1$ with the property that for every ordinal α there is a partial order $\mathbb{P}$ satisfying the following statements.*

1. *$\mathbb{P}$ is σ-distributive and satisfies the $\aleph_2$-chain condition.*

2. *$\mathbb{1}_{\mathbb{P}} \Vdash_{\mathbb{P}}$ " $\tau(\check{G}) \geq \check{\alpha}$ ".*

In the above situation, we say that G has *unbounded potential automorphism tower heights.* The following section contains the proof of this result. We close this introduction with questions motivated by this result.

Question 1.9. *Is L a model of the statement of Theorem 1.8?*

Given a partial order $\mathbb{P}$, consider the following property.

(∗) $\mathbb{P}$ is σ-closed and satisfies the $\aleph_2$-chain condition.

Question 1.10. *Is it consistent with the axioms of ZFC that there is a centreless group G of cardinality $\aleph_1$ with the property that for every ordinal α there is a partial order $\mathbb{P}$ that satisfies (∗) and $\mathbb{1}_{\mathbb{P}} \Vdash_{\mathbb{P}}$ " $\tau(\check{G}) \geq \check{\alpha}$ "?*

If this question has a positive answer, then it is natural to ask the following question.

Question 1.11. *Is it provable in ZFC that there is a partial order $\mathbb{P}$ that satisfies (∗) and forces the statement of Question 1.10 to hold true in every $\mathbb{P}$-generic extension of the ground model?*

A negative answer to this questions opens the possibility of finding a solution to Problem 1.5 using an iterated forcing construction. This solution would require an analysis of the absoluteness properties of the long automorphism towers added by Theorem 1.4 under σ-closed forcing and the use of techniques similar to the ones used in the next section.

2 Constructing the Model

In this section, we construct a ZFC-model containing a group with unbounded potential automorphism tower heights. This construction is motivated by Joel David Hamkins' alternative proof of the consistency of the *Maximality Principle* (see Hamkins 2003, page 533). We combine those techniques with Theorem 1.4 and some basic forcing results. For completeness, we provide proofs of these results.

Let $\mathcal{L}_{\in,\dot{\delta}}$ be the first-order language extending the language $\mathcal{L}_\in$ of set theory by a constant symbol $\dot{\delta}$. Given an $\mathcal{L}_\in$-formula $\varphi(v_0,\ldots,v_{n-1})$, we let $\varphi^{V_{\dot{\delta}}}(v_0,\ldots,v_{n-1})$ denote the $\mathcal{L}_{\in,\dot{\delta}}$-formula

$$(\exists x) \left[\varphi^x(v_0,\ldots,v_{n-1}) \wedge (\forall y)\left(y \in x \leftrightarrow \text{rnk}(y) < \dot{\delta}\right)\right],$$

where $\varphi^x(v_0,\ldots,v_{n-1})$ is the usual relativization of $\varphi(v_0,\ldots,v_{n-1})$ to x. We let "$V_{\dot{\delta}} \prec V$" denote the $\mathcal{L}_{\in,\dot{\delta}}$ - theory of all axioms of the form

$$(\forall x_0,\ldots,x_{n-1}) \, [(\text{rnk}(x_0),\ldots,\text{rnk}(x_{n-1}) < \dot{\delta} \wedge \varphi(x_0,\ldots,x_{n-1}))$$
$$\longrightarrow \varphi^{V_{\dot{\delta}}}(x_0,\ldots,x_{n-1})],$$

where $\varphi(v_0,\ldots,v_{n-1})$ is an n-ary $\mathcal{L}_\in$-formula.

Proposition 2.1. *If* ZF *is consistent, then so is the theory* ZFC $+$ (GCH) $+$ "$V_{\dot{\delta}} \prec V$ ".

Proof. The consistency of ZF implies the existence of a model $\langle V, \in \rangle$ of ZFC+(GCH). Given a finite fragment F of "$V_{\dot{\delta}} \prec V$ ", we can apply Levy's Reflection Principle (see Moschovakis 2009, Theorem 8C.4) to find an element δ of V with

$$\langle V, \in, \delta \rangle \models F + \text{``}\dot{\delta} \text{ is a strong limit cardinal''}.$$

By the Compactness Theorem, this shows that the theory ZFC $+$ (GCH) $+$ "$V_{\dot{\delta}} \prec V$ " is consistent. □

Proposition 2.2. *If* $\langle V, \in, \delta \rangle$ *is a model of* ZFC $+$ "$V_{\dot{\delta}} \prec V$ ", *then δ is a strong limit cardinal in* V.

Proof. By elementarity, we have $\langle V_\delta, \in \rangle \models$ (Inf) and this implies $\omega < \delta$. Further applications of elementarity yield $2^\kappa < \delta$ for every cardinal $\kappa < \delta$. □

If $V[G]$ is a generic extension of the ground model and α is an ordinal, then we also write $V[G]_\alpha$ to denote the set of all elements x of $V[G]$ with $\text{rnk}(x)^{V[G]} < \alpha$. Given a model $\langle V, \in, \delta \rangle$ of ZFC $+$ "$V_{\dot{\delta}} \prec V$ ", we investigate forcing extensions of $\langle V, \in \rangle$ by partial orders contained in V_δ.

Proposition 2.3. *Let δ be a cardinal and $\mathbb{P} \in V_\delta$ be a partial order. If G is $\mathbb{P}$-generic over* V, *then*

$$V[G]_\delta = \{\dot{x}^G \in V[G] \mid \dot{x} \in V^\mathbb{P} \cap V_\delta\}.$$

Proof. Let G be $\mathbb{P}$-generic over V. An easy induction shows that $\mathrm{rnk}(\dot{x}^G) \leq \mathrm{rnk}(\dot{x})$ holds for each name $\dot{x} \in V^{\mathbb{P}}$.

We define a sequence $\langle \dot{v}_\alpha \in V^{\mathbb{P}} \mid \alpha < \delta \rangle$ of names in the following way.

1. $\dot{v}_0 = \emptyset$.

2. For all $\alpha < \delta$, we define $\dot{v}_{\alpha+1} = \mathcal{P}\,(\mathrm{dom}(\dot{v}_\alpha) \times \mathbb{P}) \times \{\mathbb{1}_{\mathbb{P}}\}$.

3. If $\alpha \in \mathrm{Lim} \cap \delta$, then we define $\dot{v}_\alpha = \bigcup\{\dot{v}_\beta \mid \beta < \alpha\}$.

If $\mathrm{rnk}(\mathbb{P}) = \alpha_0 < \delta$, then another easy induction shows that $\mathrm{rnk}(\dot{v}_\alpha) < \alpha_0 + \alpha + \omega$ and

$$\mathbb{1}_{\mathbb{P}} \Vdash_{\mathbb{P}} \text{ "} (\forall x)\, [\mathrm{rnk}(x) < \check{\alpha} \longrightarrow x \in \dot{v}_\alpha] \text{ "}$$

holds for all $\alpha < \delta$ (the proof of Kunen 1980, Theorem 4.2, page 201 contains the details of the successor case).

Now, suppose $\dot{x} \in V^{\mathbb{P}}$ with $\mathrm{rnk}(\dot{x}^G)^{V[G]} < \delta$. By the above constructions, there is an $\alpha < \delta$ and a condition $p \in G$ with $p \Vdash_{\mathbb{P}} \text{ "} \dot{x} \subseteq \dot{v}_\alpha \text{ "}$. If we define

$$\dot{y} = \{\langle \dot{z}, r \rangle \in V^{\mathbb{P}} \times \mathbb{P} \mid (\exists q \in \mathbb{P})\, [r \leq_{\mathbb{P}} p, q \wedge \langle \dot{z}, q \rangle \in \dot{v}_\alpha \wedge r \Vdash_{\mathbb{P}} \text{ "} \dot{z} \in \dot{x} \text{"}]\} \in V^{\mathbb{P}},$$

then $\mathrm{rnk}(\dot{y}) < \delta$ and $p \Vdash_{\mathbb{P}} \text{ "} \dot{x} = \dot{y} \text{"}$. □

Lemma 2.4. *Let $\langle V, \in, \delta \rangle$ be a model of ZFC + "$V_\delta \prec V$" and $\mathbb{P} \in V_\delta$ be a partial order. If G is $\mathbb{P}$-generic over V, then $\langle V[G], \in, \delta \rangle$ is a model of ZFC + "$V_\delta \prec V$".*

Proof. By our assumptions, $\langle V_\delta, \in \rangle$ is a transitive ZFC-model, G is $\mathbb{P}$-generic over V_δ, $G \in V[G]_\delta$, $V_\delta^{\mathbb{P}} = V^{\mathbb{P}} \cap V_\delta$ and Proposition 2.3 shows $V_\delta[G] = V[G]_\delta$.

Let $\varphi(v_0, \ldots, v_{n-1})$ be an n-ary $\mathcal{L}_\in$-formula and $x_0, \ldots, x_{n-1} \in V[G]_\delta$. Fix $\dot{x}_0, \ldots, \dot{x}_{n-1} \in V_\delta^{\mathbb{P}}$ with $x_i = \dot{x}_i^G$ for all $i < n$. If $\langle V[G], \in \rangle \models \varphi(x_0, \ldots, x_{n-1})$, then there is a condition $p \in G$ with $\langle V, \in \rangle \models [p \Vdash_{\mathbb{P}} \varphi(\dot{x}_0, \ldots, \dot{x}_{n-1})]$. All parameters of this statement are elements of V_δ and we can conclude $\langle V_\delta, \in \rangle \models [p \Vdash_{\mathbb{P}} \varphi(\dot{x}_0, \ldots, \dot{x}_{n-1})]$. An application of the Forcing Theorem in V_δ gives us $\langle V[G]_\delta, \in \rangle \models \varphi(x_0, \ldots, x_{n-1})$. □

In the following, we state and prove standard results that show how a generic extension of the ground model by a *big* partial order can be factored into a two-step iteration with the property that the intermediate model contains a certain *small* set from the original forcing extension and is a generic extension of the ground model by a *small* partial order. Given a boolean algebra $\mathbb{B}$, we let $\mathbb{B}^*$ denote the partial order with domain $\mathbb{B} \setminus \{0_{\mathbb{B}}\}$ ordered by the restriction of $\leq_{\mathbb{B}}$ to this set.

Lemma 2.5. *Let κ be an infinite cardinal, $\mathbb{B}$ be a complete boolean algebra and $\dot{x} \in V^{\mathbb{B}^*}$ with $\mathbb{1}_{\mathbb{B}} \Vdash_{\mathbb{B}^*} \text{ "} \dot{x} \subseteq \check{\kappa} \text{ "}$. Then there is a κ-generated complete subalgebra $\mathbb{C}$ of $\mathbb{B}$ in V and names $\dot{D}, \dot{y} \in V^{\mathbb{C}^*}$ with the following properties.*

1. *$\mathbb{1}_{\mathbb{C}} \Vdash_{\mathbb{C}^*} \text{ "} \dot{D}$ is a partial order " and there is a dense embedding $i \colon \mathbb{B}^* \longrightarrow \mathbb{C}^* * \dot{D}$ such that $i(c) = \langle c, \dot{c} \rangle$ with $c \Vdash_{\mathbb{C}^*} \text{ "} \dot{c} = \check{c} \text{ "}$ for all $c \in \mathbb{C}^*$.*

2. *If $G_0 * G_1$ is $(\mathbb{C}^* * \dot{D})$-generic over V and G is the preimage of $G_0 * G_1$ under i, then $\dot{x}^G = \dot{y}^{G_0} \in V[G_0]$.*

Proof. Given $\alpha < \kappa$, set $\mathbb{B}_\alpha = \{b \in \mathbb{B}^* \mid b \Vdash_{\mathbb{B}^*} \text{``} \check{\alpha} \in \dot{x} \text{''}\}$ and $b_\alpha = \sup_\mathbb{B} \mathbb{B}_\alpha$. Let $\mathbb{C}$ be the complete subalgebra of $\mathbb{B}$ generated by the set $\{b_\alpha \mid \alpha < \kappa\}$ and define

$$\dot{y} = \{\langle \check{\alpha}, b_\alpha \rangle \in V^{\mathbb{C}^*} \times \mathbb{C}^* \mid \alpha < \kappa, \ b_\alpha \neq 0_\mathbb{B}\} \in V^{\mathbb{C}^*}.$$

Let G be $\mathbb{B}^*$-generic over V. If $\alpha \in \dot{x}^G$, then there is a $b \in G$ with $b \Vdash_{\mathbb{B}^*} \text{``} \check{\alpha} \in \dot{x} \text{''}$ and this shows $b_\alpha \in G$ and $\alpha \in \dot{y}^{G \cap \mathbb{C}^*}$. The other direction follows directly from the fact that $b_\alpha \in \mathbb{B}_\alpha$ holds for all $\alpha < \kappa$.

There is a canonical $\mathbb{C}^*$-name $\dot{\mathbb{D}}$ with the property that, whenever G_0 is $\mathbb{C}^*$-generic over V, then $\dot{\mathbb{D}}^{G_0}$ is the partial order whose domain is the set

$$\{b \in \mathbb{B}^* \mid (\forall c \in G_0) \ b \parallel_{\mathbb{B}^*} c\}$$

ordered by the restriction of $\leq_\mathbb{B}$ to this domain.

If $b \in \mathbb{B}^*$ and G is $\mathbb{B}^*$-generic over V with $b \in G$, then $b \in \dot{\mathbb{D}}^{G \cap \mathbb{C}^*}$ and there is a $c \in G \cap \mathbb{C}^*$ with $c \Vdash_{\mathbb{C}^*} \text{``} \check{b} \in \dot{\mathbb{D}} \text{''}$. This shows that the function

$$i_0 : \mathbb{B}^* \longrightarrow \mathbb{C}^* : b \mapsto \sup_\mathbb{B}\{c \in \mathbb{C}^* \mid c \Vdash_{\mathbb{C}^*} \text{``} \check{b} \in \dot{\mathbb{D}} \text{''}\}$$

is well-defined. Pick a function $i_1 : \mathbb{B}^* \longrightarrow V^{\mathbb{C}^*}$ with $\mathbb{1}_\mathbb{C} \Vdash_{\mathbb{C}^*} \text{``} i_1(b) \in \dot{\mathbb{D}} \text{''}$ and $i_0(b) \Vdash_{\mathbb{C}^*}$ $\text{``} \check{b} = i_1(b) \text{''}$ for all $b \in \mathbb{B}^*$. Define $i : \mathbb{B}^* \longrightarrow \mathbb{C}^* * \dot{\mathbb{D}}$ by setting $i(b) = \langle i_0(b), i_1(b) \rangle$.

Given $c, c' \in \mathbb{C}^*$, it is easy to see that $c' \Vdash_{\mathbb{C}^*} \text{``} \check{c} \in \dot{\mathbb{D}} \text{''}$ is equivalent to $c' \leq_\mathbb{C} c$. This shows that $i_0(c) = c$ holds for all $c \in \mathbb{C}^*$. We show that i is a dense embedding.

Let $b_0, b_1 \in \mathbb{B}^*$ with $b_0 \leq_\mathbb{B} b_1$. Given $c \in \mathbb{C}^*$, if $c \Vdash_{\mathbb{C}^*} \text{``} \check{b}_0 \in \dot{\mathbb{D}} \text{''}$, then $c \Vdash_{\mathbb{C}^*}$ $\text{``} \check{b}_1 \in \dot{\mathbb{D}} \text{''}$. This shows that $i_0(b_0) \leq_{\mathbb{C}^* * \dot{\mathbb{D}}} i_0(b_1)$ holds and hence $i(b_0) \leq_{\mathbb{C}^* * \dot{\mathbb{D}}} i(b_1)$. Next, fix $a_0, a_1 \in \mathbb{B}^*$ with $a_0 \perp_{\mathbb{B}^*} a_1$. Assume, toward a contradiction, that there is a $\langle c, \dot{d} \rangle \in \mathbb{C}^* * \dot{\mathbb{D}}$ with $\langle c, \dot{d} \rangle \leq_{\mathbb{C}^* * \dot{\mathbb{D}}} i(a_0), i(a_1)$. We can find a $0_\mathbb{C} <_\mathbb{C} c_* \leq_\mathbb{C} c$ and a condition $d \in \mathbb{B}^*$ with $c_* \Vdash_{\mathbb{C}^*} \text{``} \check{d} = \dot{d} \text{''}$. This means $c_* \Vdash_{\mathbb{C}^*} \text{``} \check{d} \leq_{\dot{\mathbb{D}}} \check{a}_0, \check{a}_1 \text{''}$ and therefore $0_\mathbb{B} <_\mathbb{B} d \leq_\mathbb{B} a_0, a_1$, a contradiction. Finally, fix $\langle c, \dot{d} \rangle \in \mathbb{C}^* * \dot{\mathbb{D}}$. As above, there are $0_\mathbb{C} <_\mathbb{C} c_* \leq_\mathbb{C} c$ and $d \in \mathbb{B}^*$ with $c_* \Vdash_{\mathbb{C}^*} \text{``} \check{d} = \dot{d} \text{''}$. Since $c_* \Vdash_{\mathbb{C}^*} \text{``} \check{c}_* \parallel_{\dot{\mathbb{B}}} \dot{d} \text{''}$, there is a condition $d_* \in \mathbb{B}^*$ with $d_* \leq_\mathbb{B} c_*, d$. By the above computations, $i_0(d_*) \leq_\mathbb{C} i_0(c_*) = c_* \leq_\mathbb{C} c$ and $i_0(d_*) \Vdash_{\mathbb{C}^*} \text{``} i_1(d_*) \leq_{\dot{\mathbb{D}}} \dot{d} \text{''}$. This means $i(d_*) \leq_{\mathbb{C}^* * \dot{\mathbb{D}}} \langle c, \dot{d} \rangle$ and i is a dense embedding.

If $G_0 * G_1$ is $(\mathbb{C}^* * \dot{\mathbb{D}})$-generic over V and G is the preimage of $G_0 * G_1$ under i, then both G_0 and $G \cap \mathbb{C}^*$ are $\mathbb{C}^*$-generic over V. Since $i_0 \upharpoonright \mathbb{C}^* = \text{id}_{\mathbb{C}^*}$, it follows that $G \cap \mathbb{C}^* \subseteq G_0$ and the maximality of generic filters yields $G \cap \mathbb{C}^* = G_0$. By the above calculations, $\dot{x}^G = \dot{y}^{G \cap \mathbb{C}^*} = \dot{y}^{G_0} \in V[G_0]$. $\square$

Given an infinite cardinal κ and a set X, we let $[X]^{<\kappa}$ denote the set of all subsets of X of cardinality less than κ. If $\mathbb{B}$ is a boolean algebra and $\mathbb{C}$ is a subalgebra of $\mathbb{B}$, then $\mathbb{C}$ is called $<\kappa$-complete in $\mathbb{B}$ if $\inf_\mathbb{B} X \in \mathbb{C}$ for all $X \in [\mathbb{C}]^{<\kappa}$.

Proposition 2.6. *Let κ be an infinite regular cardinal, $\mathbb{B}$ be a complete boolean algebra that satisfies the κ-chain condition and $\mathbb{C}$ be a subalgebra of $\mathbb{B}$. If $\mathbb{C}$ is $<\kappa$-complete in $\mathbb{B}$, then $\mathbb{C}$ is a complete subalgebra of $\mathbb{B}$.*

Proof. Assume, toward a contradiction, that $\mathbb{C}$ is not a complete subalgebra of $\mathbb{B}$ and let ν be the least cardinal such that there is a sequence $\langle c_\alpha \in \mathbb{C} \mid \alpha < \nu \rangle$ with $\inf_\mathbb{B}\{c_\alpha \mid \alpha < \nu\} \notin \mathbb{C}$. By our assumption, $\nu \geq \kappa$ and it is easy to see that ν is a regular cardinal. Given $\alpha < \nu$, we define $b_\alpha = \inf_\mathbb{B}\{c_\beta \mid \beta < \alpha\}$. Our assumptions imply $0_\mathbb{B} \neq b_\alpha \in \mathbb{C}$ and $b_\beta \leq_\mathbb{B} b_\alpha$ for all $\alpha \leq \beta < \nu$. Moreover, $\inf_\mathbb{B}\{b_\alpha \mid \alpha < \nu\} = \inf_\mathbb{B}\{c_\alpha \mid \alpha < \nu\} \notin \mathbb{C}$. If we define $a_\alpha = b_\alpha - b_{\alpha+1}$ for all $\alpha < \nu$, then the set $A = \{a_\alpha \in \mathbb{B} \mid \alpha < \nu, a_\alpha \neq 0_\mathbb{B}\}$ is an anti-chain in $\mathbb{B}$ and has cardinality less than κ. This means that there is an $\alpha < \nu$ with $a_\beta = 0_\mathbb{B}$ for all $\alpha \leq \beta < \nu$ and an easy induction shows that this implies $b_{\alpha+1} = b_\beta$ for all $\alpha < \beta < \nu$. We can conclude $\inf_\mathbb{B}\{b_\alpha \mid \alpha < \nu\} = b_{\alpha+1} \in \mathbb{C}$, a contradiction. □

Lemma 2.7. *Let κ be an infinite cardinal, $\mathbb{B}$ be a complete boolean algebra that satisfies the κ^+-chain condition and C be a subset of $\mathbb{B}$ of cardinality at most κ. If $\mathbb{C}$ is the complete subalgebra of $\mathbb{B}$ generated by C, then $\mathbb{C}$ has cardinality at most 2^κ.*

Proof. It suffices to construct a complete subalgebra $\mathbb{C}^+$ of $\mathbb{B}$ that contains C and has cardinality at most 2^κ. We define an ascending sequence $\langle \mathbb{C}_\alpha \mid \alpha < \kappa^+ \rangle$ of subalgebras of $\mathbb{B}$ in the following way.

1. $\mathbb{C}_0$ is the subalgebra of $\mathbb{B}$ generated by C.

2. If $\alpha \in \kappa^+ \cap \mathrm{Lim}$, then $\mathbb{C}_\alpha = \bigcup\{\mathbb{C}_\beta \mid \beta < \alpha\}$.

3. $\mathbb{C}_{\alpha+1}$ is the subalgebra of $\mathbb{B}$ generated by the set $\{\inf_\mathbb{B} X \mid X \in [\mathbb{C}_\alpha]^{<\kappa^+}\}$ for all $\alpha < \kappa^+$.

An easy induction shows that the subalgebra $\mathbb{C}_\alpha$ has cardinality at most 2^κ for all $\alpha < \kappa^+$ and this shows that the subalgebra $\mathbb{C}^+ = \bigcup\{\mathbb{C}_\alpha \mid \alpha < \kappa^+\}$ also has cardinality at most 2^κ. We show that $\mathbb{C}^+$ is a complete subalgebra of $\mathbb{B}$. By Proposition 2.6, it suffices to show that $\mathbb{C}^+$ is $<\kappa^+$-complete in $\mathbb{B}$. If $X \in [\mathbb{C}^+]^{<\kappa^+}$, then there is an $\alpha < \kappa^+$ with $X \subseteq \mathbb{C}_\alpha$. But this means $\inf_\mathbb{B} X \in \mathbb{C}_{\alpha+1} \subseteq \mathbb{C}^+$. □

We are now ready to prove our main result.

Proof of Theorem 1.8. By Proposition 2.1, the consistency of ZFC gives us a model $\langle V, \in, \delta \rangle$ of ZFC + (GCH) + "$V_\delta \prec V$". Work in V and apply Theorem 1.4 with $\kappa = \omega_1$ and $\nu = \alpha = \delta^+$ to produce a partial order $\mathbb{P}$ with the above properties. By (Kunen 1980, Lemma 3.3, page 63), there is a complete boolean algebra $\mathbb{B}$ and a dense embedding $d: \mathbb{P} \longrightarrow \mathbb{B}^*$.

We can find a name $\dot{x} \in V^{\mathbb{B}^*}$ with the property that, whenever H is $\mathbb{B}^*$-generic over V, then $\dot{x}^H \subseteq \omega_1$ and there is a centreless group $G \in V[H]$ with domain ω_1, $\tau(G)^{V[H]} = \delta$ and

$$\alpha \circ_G \beta = \gamma \iff \ll\alpha,\beta\gg,\gamma> \in \dot{x}^H \tag{8.1}$$

for all $\alpha, \beta, \gamma < \omega_1$.[3] Apply Lemma 2.5 with ω_1, $\mathbb{B}$ and $\dot{x}$ to find $\mathbb{C}, \dot{\mathbb{D}}, i$ and $\dot{y}$ with the above properties. Since $\mathbb{B}$ satisfies the $\aleph_2$-chain condition, we can apply Lemma 2.7 to see that $\mathbb{C}$ has cardinality at most $2^{\aleph_1} < \delta$. Let $\mathbb{P}_0 \in V_\delta$ be a partial order isomorphic to $\mathbb{C}^*$ in V. Our construct ensures that $\mathbb{P}_0$ is σ-distributive and satisfies the $\aleph_2$-chain condition. Let F be $\mathbb{P}_0$-generic over V. By Lemma 2.4, $\langle V[F], \in, \delta \rangle$ is a model of ZFC + "$V_\delta \prec V$". We will show that $\langle V[F], \in \rangle$ contains a group with unbounded automorphism tower heights.

---

[3]We let $\ll\cdot,\cdot\gg : \mathrm{On} \times \mathrm{On} \longrightarrow \mathrm{On}$ denote the Gödel pairing function.

Let $H_0 \in V[F]$ be the image of F under the isomorphism between $\mathbb{P}_0$ and $\mathbb{C}^*$. We have $V[F] = V[H_0]$. Let H_1 be $\dot{\mathbb{D}}^{H_0}$-generic over $V[H_0]$ and H be the preimage of $H_0 * H_1$ under i. In $V[H] = V[H_0][H_1]$, there is a centreless group G with domain ω_1 and $\tau^{V[H]} = \delta$ whose group operation is coded by $\dot{x}^H$ as in (8.1). The above construction yields $\dot{x}^H \in V[H_0]$ and this means $G \in V[H_0]_\delta = V[F]_\delta$. In $V[H_0]$, $\dot{\mathbb{D}}^{H_0}$ is a σ-distributive partial order that satisfies the $\aleph_2$-chain condition and we have

$$\langle V[H_0], \in \rangle \models \left[\mathbb{1}_{\dot{\mathbb{D}}^{H_0}} \Vdash_{\dot{\mathbb{D}}^{H_0}} \text{``} \tau(\check{G}) > \check{\delta} \text{''} \right]. \tag{8.2}$$

Let $\varphi(v_0, \ldots, v_4)$ be the $\mathcal{L}_\in$-formula that is the conjunction of the following statements:

1. "v_0 is a σ-distributive partial order satisfying the $\aleph_2$-chain condition",

2. "v_1 is a centreless group and v_2 is the canonical v_0-name for v_1",

3. "v_3 is an ordinal and v_4 is the canonical v_0-name for v_3",

4. $\mathbb{1}_{v_0} \Vdash_{v_0} \text{``} \tau(v_2) \geq v_4 \text{''}$.

In order to show that G is a group with unbounded potential automorphism tower heights in $\langle V[F], \in \rangle$, it clearly suffices to show

$$\langle V[F], \in \rangle \models (\forall \alpha \in \mathrm{On})(\exists x, y, z)\, \varphi(x, G, y, \alpha, z). \tag{8.3}$$

Given $\alpha < \delta$, the above (8.2) implies

$$\langle V[F], \in \rangle \models (\exists x, y, z)\, \varphi(x, G, y, \alpha, z)$$

and all parameters of this statement are contained in $V[F]_\delta$. By elementarity, we have

$$\langle V[F]_\delta, \in \rangle \models (\forall \alpha \in \mathrm{On})(\exists x, y, z)\, \varphi(x, G, y, \alpha, z),$$

and another application of elementarity shows that (8.3) holds. □

References

Fuchs, G. and J. D. Hamkins (2008). "Changing the Heights of Automorphism Towers by Forcing with Souslin Trees over L". In: *Journal of Symbolic Logic* 73, pp. 614–633.

Fuchs, G. and P. Lücke (2011). "Iteratively Changing the Heights of Automorphism Towers". In: *Notre Dame Journal of Formal Logic (forthcoming)*.

Hamkins, J. D. (2002). "How Tall is the Automorphism Tower of a Group?" In: *Logic and Algebra*. Providence, RI: American Mathematical Society, pp. 49–57.

— (2003). "A Simple Maximality Principle". In: *Journal of Symbolic Logic* 68, pp. 527–550.

Hamkins, J. D. and S. Thomas (2000). "Changing the Heights of Automorphism Towers". In: *Annals of Pure and Applied Logic* 102, pp. 139–157.

Just, W., S. Shelah, and S. Thomas (1999). "The Automorphism Tower Problem Revisited". In: *Advances in Mathematics* 148, pp. 243–265.

Kunen, K. (1980). *Set Theory*. Studies in Logic and the Foundations of Mathematics 102. An introduction to independence proofs. Amsterdam: North-Holland.

Moschovakis, Y. N. (2009). *Descriptive Set Theory*. Providence, RI: American Mathematical Society.

Thomas, S. *The Automorphism Tower Problem*. Forthcoming.

— (1985). "The Automorphism Tower Problem". In: *Proceedings of the American Mathematical Society* 95, pp. 166–168.

— (1998). "The Automorphism Tower Problem. II". In: *Israel Journal of Mathematics* 103, pp. 93–109.

Wielandt, H. (1939). "Eine Verallgemeinerung der invarianten Untergruppen". In: *Mathematische Zeitschrift* 45, pp. 209–244.

CHAPTER 9

The Surprise Examination Paradox in Dynamic Epistemic Logic

Alexandru Marcoci

Abstract

The surprise examination paradox has been the topic of many philosophical papers. However, despite its long history, no long-lasting solution has emerged. The debates related to the surprise examination paradox go deeper than what the correct solution is. One highly debated issue that I directly address in this paper is the following: is there a solution to be given or should we embrace the surprise examination paradox as a true paradox and look for an inconsistency in our conceptions on knowledge, belief, etc.? Kaplan and Montague (1960) argue that the surprise examination paradox is indeed a paradox, while J Gerbrandy (2007) and Baltag and Smets (2010) believe that a solution could be given but not if the teacher's announcement is meant to be fulfilled (Gerbrandy), or if the students are to trust the teacher (Baltag and Smets). Of course, such approaches have a lot of merit and all authors manage to use their (negative) conclusions to argue for deep philosophical conclusions regarding knowledge and the way in which agents revise their beliefs in the face of new information. However, these approaches fail to meet widely accepted criteria that philosophers expect a solution to the surprise examination paradox to meet. They have been explicitly set up by Wright and Sudbury (1977) and contain the idea that the students should be surprised even after the teacher's announcement (which neither Kaplan and Montague, nor Gerbrandy can accommodate) and that a surprise examination is indeed possible (which Baltag and Smets cannot accept).

In this paper I show that dynamic epistemic logic (DEL) can guide us towards a philosophically informed solution to the surprise examination paradox. The question that drives the analysis is: is there a way of coming up with an intuitive solution to the surprise examination paradox that meets the criteria of Wright and Sudbury? I argue that there is indeed such an interpretation. The paper is organized as follows: in the beginning I briefly present some key concepts from dynamic epistemic logic; I then briefly present the solutions offered so far in DEL and why there is a need for yet another solution. In the following section I show how to translate different aspects of the paradox (e.g. the notion of surprise, the meaning of the teacher's announcement etc.) into the language of DEL. I argue that DEL provides a way of distinguishing between different parts of the teacher's announcement and that when (the effect of) each part (on the student) is correctly understood, a solution to the paradox that meets all the conditions of Wright and Sudbury emerges.

1 Introduction

A teacher announces that there will be a surprise examination the following week. A student objects that this is impossible: "The class meets on Monday, Wednesday, and Friday. If the examination is given on Friday, then on Thursday I would be able to predict that the examination is on Friday. It would not be a surprise. Can the examination be given on Wednesday? No, because on Tuesday I would be in a position to predict that the examination will not be on Friday (thanks to the previous reasoning) and know that the examination was not on Monday (thanks to memory). Therefore, on Tuesday I could foresee that the examination will be on Wednesday. An examination on Wednesday would not be a surprise. Could the surprise examination be on Monday? On Sunday, the previous two eliminations would be available to me. Consequently, I would be in a position to predict that the examination must be on Monday. So a Monday examination would also fail to be a surprise. Therefore, it is impossible for there to be a surprise examination."[1]

Ultimately, the teacher gives the students an examination on one of the available days, say Wednesday (note: he could give it on any other day as well!). The paradoxical flavor of this story comes in when we adopt the student's perspective: she begins by assuming that there will be a surprise examination, she deduces that such a surprise examination cannot take place, and then she is indeed surprised when the examination actually comes.

The surprise examination paradox has been the topic of many philosophical papers. However, despite its long history, no long-lasting solution emerged. The debates related to the surprise examination paradox go deeper than what the correct solution is. One highly debated issue that I directly address in this paper is the following: is there a solution to be given or should we embrace the surprise examination paradox as a true paradox and look for an inconsistency in our conceptions on knowledge, belief, etc.? Kaplan and Montague (1960) argue that the surprise examination paradox is indeed a paradox, while J Gerbrandy (2007) and Baltag and Smets (2010) believe that a solution could be given but not if the teacher's announcement is meant to be fulfilled (Gerbrandy), or if the students are to trust the teacher (Baltag and Smets). Of course,

---

[1] Formulation adapted from Sorensen (2009).

such approaches have a lot of merit and all authors manage to use their (negative) conclusions to argue for deep philosophical conclusions regarding knowledge and the way in which agents revise their beliefs in face of new information. However, these approaches fail to meet widely accepted criteria that philosophers expect a solution to the surprise examination paradox to meet. They have been explicitly set up by Wright and Sudbury (1977) and contain the idea that the students should be surprised even after the teacher's announcement (which neither Kaplan and Montague, nor Gerbrandy can accommodate) and that a surprise examination is indeed possible (which Baltag and Smets cannot accept).

In this paper I show that dynamic epistemic logic (DEL) can guide us towards a philosophically informed solution to the surprise examination paradox. The question that drives the analysis is: is there a way of coming up with an intuitive solution to the surprise examination paradox that meets the criteria of Wright and Sudbury? I argue that there is indeed such an interpretation. The paper is organized as follows: in the beginning I briefly present some key concepts from dynamic epistemic logic; I then briefly present the solution offered so far in DEL and why there is a need for yet another solution. In the following section I show how to translate different aspects of the paradox (e.g. the notion of surprise, the meaning of the teacher's announcement etc.) into the language of DEL. I argue that DEL provides a way of distinguishing between different parts of the teacher's announcement and that when (the effect of) each part (on the student) is correctly understood, a solution to the paradox that meets all the conditions of Wright and Sudbury emerges.

2 Three Types of Information: Hard, Soft, Procedural

DEL has recently suggested a distinction between three different kinds of information: hard, soft and procedural. The first two types of information are "epistemic information about the facts in the world and what others know" (van Benthem 2011, p. 62), while the third type of information, i.e. procedural, is information about how the world will change: "about the process generating the epistemic information" (van Benthem 2011, p. 62).[2] van Benthem (2006) defines the hard information an agent has as the set of possible worlds between which that agent cannot distinguish. Hence, receiving hard information affects the range of possible worlds an agent entertains as possible. In the same manner the soft information an agent has could be defined as his set of plausibility relations. Consequently, receiving soft information affects the plausibility relations between the states that the agent entertains as possible. To wit, hard information changes what I know, while soft information changes what I believe.

There have been logics developed for describing the way in which hard and soft information flows from one agent to another. (i) A way of describing hard information

---

[2]The notions of hard and soft information have received much attention in the philosophical literature. Even if they have not been referred to in this way, knowledge and belief and the ways in which they come about and are revised have always been at the center of epistemological concerns. However, philosophers have so far neglected procedural (or protocol) information. Schwitzgebel (2010) does mention procedural knowledge, but what he means by it is 'skills gained unconsciously'. That is, autistic patients who are exposed to the same task every day improve their performances with the passing of time, however they have no memories about doing that task. Hence, it cannot be said that they know, in the declarative sense, how to solve the task, but they do know, in the practical sense, how to solve it. But this is not what procedural information does. Procedural information gives an agent declarative knowledge about future actions that himself or others will do.

flow is by extending the language of epistemic logic with a 'dynamic operator' $[!\varphi]\psi$. Its intended interpretation is: 'ψ holds after a public and truthful announcement of φ has been made.' This extended language allows for (finite) iterations of both the knowledge operator and the announcement operator and mixes in between, that is: $K_i K_j \varphi$ and $[!\varphi][!\psi]\chi$ as well as $[!\varphi]K_i\psi$ and $K_i[!\varphi]\psi$ are well-defined. The semantics of the new operator is given by the following clause:

$$M, s \models [!\varphi]\psi \text{ iff } M, s \models \varphi \Rightarrow M^{!\varphi}, s \models \psi.$$

Given an epistemic model $M = \langle S, R_i, V \rangle$ and a sentence $\varphi \in \mathcal{L}_{EL}$, the updated model, $M^{!\varphi} = \langle S^{!\varphi}, R_i^{!\varphi}, V^{!\varphi} \rangle$, is defined in the following manner:

$$S^{!\varphi} = \{u \in S : M, u \models \varphi\},$$
$$R_i^{!\varphi} = R_i \cap (S^{!\varphi} \times S^{!\varphi}),$$
$$V^{!\varphi} = V.$$

Remark that the dynamic nature of the new operator is clear from its semantics: the statement that $[!\varphi]\psi$ is interpreted at model M, by reference to a different model $M^{!\varphi}$. Moreover, the effect of the public announcement of sentence φ on an epistemic model M is the elimination of all states in M at which φ did not hold. This logic is known as public announcement logic (PAL).

(ii) A way of describing soft information is by extending the language of epistemic logic with a dynamic operator $[\Uparrow \varphi]\psi$. Its intended meaning is 'after a lexicographic upgrade with φ, ψ is the case.' After a lexicographic upgrade with sentence φ, all the φ-states become more plausible than the $\neg\varphi$- states, while the order within the $(\neg)\varphi$-states remains unchanged.[3] Moreover, the epistemic relations between the states remain unchanged. Thus the operator described only affects the doxastic relations between states and as such only changes the beliefs and not the knowledge. Of course, beliefs could be changed in many ways and, accordingly, there exist a host of different operators to describe all these soft announcements. The different ways of revising one's beliefs and their characteristics are investigated in dynamic belief revision theory.

Procedural information is information about what the agents can do or about what can happen to them. Following Pacuit and Simon (2011) we can think of the procedural information an agent has as a "subtree from the *grand stage* of all possible sequence of events that could take place in interactive situations" (p. 2). Hard information affects the knowledge of an agent, soft information affects the beliefs; procedural information affects the protocol that an agent knows or believes to be running. A

---

[3] In order to interpret belief and the effects of soft information on it, the underlying model is best construed as a plausibility model. Given a (finite) set of agents, *Agt*, with $i, j, k, \ldots$ as its elements, and a countable set of atoms *Prop*, with $p, q, r, \ldots$ as its elements, a plausibility model is a structure $M = \langle S, \leq_i, V \rangle$, where:

1. S is a (finite) non-empty set of states;
2. $\leq_i$ is a set of reflexive, transitive, binary relations on S, such that every non-empty subset has maximal elements;
3. V assigns a set of states to each propositional variable from the set *Prop*, that is $V: Prop \rightarrow 2^S$.

In such a model, a sentence is believed if it is the case in the most plausible states, while a sentence is known at a state s if for every t such that either $s \leq_i t$ or $t \leq_i s$, that sentence is the case in t.

protocol is a description of how the world can evolve in the future.[4] In this sense, the protocol can encode: physical possibilities, i.e. in a world in which I have not bought a lottery ticket, winning the lottery cannot be in the protocol; an agent's attitudes and plans, i.e. if an agent a trusts another agent b completely, b's protocol should be so that whenever he learns what a believes he revises his beliefs in order to match a's; certain conventions, i.e. if we decided that the proper way of starting a conversation is by saluting, then immediately after we meet, the only action available to both of us is to salute; or the path not taken - if an event e occurred, the way in which I will change my attitude towards it depends on what other events could have happened, i.e. your attitude towards receiving 10 dollars can vary according to how much more money you could have made and at what cost.[5]

Formally, the protocol can be defined thus. Given a non-empty set of events Σ, a history, h, on Σ is a finite sequence of events from Σ. Let Σ^* be a set of histories on Σ. Then, a protocol on Σ is a set $H \subseteq \Sigma^*$ closed under non-empty finite prefixes. This construction, then, can be assigned to an entire domain of a model, that is, at every state the same protocol is in place, or it can be assigned to individual states of the model. Also, this protocol generates from a plausibility/epistemic model a temporal structure (tree-like) on which more operators could be defined. For instance, (i) $NEXT\varphi$ which is the case if and only if $\exists e$ such that $he \in H$ and after he has happened, φ is the case; (ii) $FUTURE\varphi$ which is the case if and only if $\exists\sigma$ such that $h\sigma \in H$ and after $h\sigma$ has happened, φ is the case; and (iii) $BEFORE\varphi$ which is the case if and only if $\exists e, \exists h'$ such that $h = h'e$ and after h' has happened φ is the case.

3 The Surprise Examination Paradox in DEL: the Road so Far

I am not the first to notice how suited DEL seems to be for solving the surprise examination paradox: (i) J. Gerbrandy (1999), van Ditmarsch and Kooi (2006) and J Gerbrandy (2007), on the one hand and (ii) Baltag and Smets (2010), Baltag (2009) and Baltag (2010), on the other, believe that a solution to the surprise examination paradox could be given if we analyze the facts with the right tools. These are public announcement logic according to Gerbrandy and dynamic belief revision according to Baltag and Smets.[6]

Gerbrandy's solution follows the tradition started by Binkley (1968) and continued by Wright and Sudbury (1977) of seeing the surprise examination paradox as stemming from Moore's paradox (Moore 1993). The conclusion of Gerbrandy's analysis is that the teacher's announcement, just like a Moorean sentence, can be announced but

---

[4]Intuitively, epistemic and procedural information are obviously distinct. A nice feature of DEL, apart from being able to distinguish formally between these two types of information is that it can also provide a formal way of expressing their distinctness. DEL without protocols usually reduces to the basic epistemic logic via a set of reduction axioms. However, dynamic epistemic logic with protocols cannot be reduced to epistemic logic. This shows that procedural information is a distinct type of information from epistemic one, and the formal framework for dealing with epistemic information cannot accommodate procedural information.

[5]This last idea has not yet been investigated in DEL. However, it has received some attention in probability theory, see Shafer (1985).

[6]In recent years, DEL has become the general name for a large family of different logics that are all inspired by the idea of 'a dynamic turn in logic'. Therefore, in this sense, DEL includes public announcement logic, dynamic belief revision, temporal public announcement logic, temporal dynamic epistemic logic etc. However DEL can also refer to a particular logic developed in Baltag, Moss, and Solecki (1998). I will always use DEL to refer to its more general meaning.

it cannot come to be known. This is a well-known phenomenon in dynamic epistemic logic, where it is usually phrased as: 'the teacher's announcement is not *successful*'. A sentence not being successful means that it cannot come to be known after being truthfully uttered/written; think of 'it's raining but you don't believe that it is raining': if the person hearing/reading it trusts the person uttering/writing it, he will immediately recognize that the sentence is true, but will not be able to come to know it himself. Just like Moore's sentence, the surprise sentence, as Gerbrandy construes it, cannot become known after being uttered. Therefore, it will not be the case that the students will be surprised whenever the examination comes. Gerbrandy concludes that when the surprise examination paradox is interpreted by means of public announcement logic it becomes obvious that it is not a paradox: the teacher's announcement is not a contradiction, but just a sentence which cannot become known to the students. Moreover, the error in the students initial reasoning can now be clearly stated: it lies in assuming that the teacher's announcement is successful, which under scrutiny, was proven to be wrong.

Baltag and Smets continue the tradition of Quine (1953) and interpret the surprise examination paradox as a problem involving the interpretation of *trust*. Their conclusion is that the paradox withers away once one realizes that there is no reason for the students to continue trusting the teacher after they perform the backward elimination argument. Therefore, after they reach contradiction (which Baltag and Smets argue that does happen as long as the students trust the teacher), the students revise their attitude towards the teacher until they can integrate what the teacher announces with the rest of their beliefs. The mechanism of belief revision has been intensively studied in DEL and Baltag and Smets prove, by the mechanism provided by DEL that there is an attitude that the students can consistently entertain towards the teacher, and what is more, it is unique. They call it the *minimal AGM[7] norm* and they define it as: 'φ comes to be believed after it is announced unless, after the announcement, it comes to be known that it's false'. Baltag and Smets argue that the students can have this attitude towards the teacher because they come to know that the teacher's announcement is false. In consequence, the students will know the teacher lied, and the exam will not be a surprise. Therefore, they will expect it, whenever it comes, even on Monday!

Despite their considerable appeal,[8] neither Gerbrandy's nor Baltag and Smets' solutions meet some very plausible intuitions about the surprise examination paradox. The first to explicitly state the conditions that a solution to the surprise examination has to meet in order to be an intuitively satisfying solution were Wright and Sudbury (1977):

1. The solution should make the teacher's announcement satisfiable.[9]

2. The solution should make it clear that the teacher can carry out the announcement even after he has announced it.[10]

3. The solution should do justice to the intuitive meaning of the announcement.

---

[7]These initials refer to the names of the three authors of a seminal work on belief revision theory, namely Alchourrón, Gärdenfors, and Makinson (1985).

[8]To which the formal arguments supporting their conclusions play a significant role.

[9]"since a surprise examination is, palpably, a logical possibility" (Wright and Sudbury 1977).

[10]"since, palpably, he can." (Wright and Sudbury 1977).

4. The solution should do justice to the intuitive plausibility of the pupils' reasoning.

5. The solution should make it possible for the students to be informed by the announcement.[11]

6. The solution should explain the role, in the generation of the puzzle, of the announcement's being made to the students.[12,13]

Let us now return to the DEL solutions proposed so far: firstly, J Gerbrandy (2007) argues that his solution meets the conditions of Wright and Sudbury (1977). However, I believe Gerbrandy fails to do so, at least in the case of conditions (2) and (3). The reason why Gerbrandy's solution fails to meet condition (2) is the following: since the teacher's announcement is not successful, the students will not know after the announcement that they will be surprised. This means that after the announcement, there is at least one situation the students consider as possible in which, if the examination comes, the students will have expected it. Therefore, it is not the case that the teacher's announcement can be carried out. More precisely, there are indeed cases in which if the examination comes, the students will be surprised even after the teacher's announcement, but it is not the case that whenever the examination comes the students will be surprised. If the latter would have been the case, then the teacher's announcement would have been successful. However, the second condition asks for the students to be surprised, whenever the examination comes, and indeed, Wright and Sudbury's solution tries to meet this condition. Thus, Gerbrandy's solution fails to meet condition (2). Moreover, Gerbrandy's solution also fails to meet condition (3). The reason is the following: Gerbrandy interprets the teacher's announcement as referring to what is the case before he makes the announcement. In other words, what the teacher announces, according to Gerbrandy, is that "this week's examination will be a surprise!", not that "This week's examination will be a surprise and will remain a surprise even after learning *this*!". However, as almost all authors since the seminal contribution of Shaw (1958) have claimed, the teacher's announcement has to be self-referential: it has to state that even after the students learn it, the examination will still come as a surprise. Thus, Gerbrandy's solution fails to meet condition (3).

Secondly, Baltag and Smets fail to meet conditions (2) and (4). The motivation for why they fail to meet condition (2) is straightforward: since after the teacher's announcement the students know he has lied to them and since they then expect the examination whenever this is given, it is not the case that the teacher can carry out his announcement. The motivation for why they fail to meet condition (4) is equally

---

[11] Sorensen (1988) interprets this condition as saying that the students should be able to come to know what the teacher announced: "if you cannot trust your teachers, who can you trust?" (p. 312).

[12] "there is, intuitively, no difficulty if, e.g. the teacher tells only the second teacher or keeps his intentions to himself."

[13] Other authors adhere implicitly or explicitly to this set of conditions. For example J Gerbrandy (2007) explicitly claims that his theory meets these conditions (although I will shortly argue that this is not the case). Other authors such as Sorensen (1988) and Williamson (2000) offer solutions that are in the spirit of these conditions, although no reference to them is explicitly made. All in all, Wright and Sudbury's conditions are usually accepted by most and I believe that these conditions are indeed necessary for a solution to the surprise examination paradox in order for it to be intuitively satisfactory. However, I do not agree with Wright and Sudbury's claim that they are also sufficient. Nevertheless, from now on, because I believe they are necessary I will treat them as a testbed for any solution to the surprise examination paradox.

straightforward. Baltag and Smets claim that in order to avoid the paradox the students have to revise their trust towards the teacher. But this explains the student's error by them not realizing that they can revise their attitude towards the teacher. I believe, however, that such an answer makes the students too naive (more than intuitively acceptable!).

Therefore, so far the DEL solutions to the surprise examination paradox seem to suggest that a surprise examination cannot be announced with the intention of it actually being a surprise (whenever it comes)! However, I believe that DEL can also help us come up with a more positive solution to the surprise examination paradox, which accommodates our intuitions (by which I mean Wright and Sudbury's) regarding the scenario. In consequence, the question that drives the investigation in this essay is: 'can DEL vindicate our pre-theoretical intuitions regarding the surprise examination paradox?'.

4 The Surprise Examination Paradox in DEL: a New Solution

Surprise appears to be an attitude that always accompanies an act of learning. If you sit on your couch and you see nothing new, or you make no inferences, etc. you cannot get surprised. Moreover, if the new information that you receive is something that you were considering possible beforehand, this, again, does not seem to be surprising: if you are sitting on the couch and are looking for the remote, which you believe that it is either on the table or on the TV, then seeing it on the TV will not surprise you. However, seeing it in the microwave oven would. Therefore, I believe that surprise is best construed as accompanying an act of learning an information that you previously believed to be impossible.

So I propose as a starting point the following definition of surprise: $B\varphi \land BEFORE(B\neg\varphi)$. This reads: 'you now believe that φ but at the last step you believed that φ cannot be the case'. The state at which this sentence is true is the state at which you learned that φ. The *BEFORE* operator allows us to identify the state at which the act of learning occurred. Moreover, this is a special type of learning: you learned that something is the case, while before you believed it to be impossible. I will adopt this definition of surprise for now, and I will review it later.

Let us return to what happens in the surprise examination paradox. First, the student is indifferent between there being an examination in the following week and not. That is,

$$M: \quad \boxed{s_1 : mo} \leftarrow - \rightarrow \boxed{s_2 : we} \leftarrow - \rightarrow \boxed{s_3 : fr} \leftarrow - \rightarrow \cdots$$

The meaning of mo, we and fr is 'the examination is scheduled for Monday', 'the examination is scheduled for Wednesday', and respectively 'the examination is scheduled for Friday', while the arrows represent the plausibility relations for the student. The teacher makes his announcement: 'next week you will receive a surprise examination!' Although apparently very simple, I believe there are three distinct parts to this announcement. The first part of the announcement (A) is an existential claim: one of mo, we and fr is the case. Many authors have noted that it makes a lot of sense to stipulate that the student trusts the teacher. For example, Sorensen (1988) asks: "if you cannot trust your teachers, who can you trust?" (p. 312). Hence from this point onwards I will assume that the student takes the teacher as being a highly trusted source. So after the announcement of part (A), the student will try to revise her beliefs in such a way so that she comes to believe that (A). That is, the student has to believe that there will be an exam the following week. This means that (A) is a public announcement of the fact that one of *mo*, *we* and *fr* has to be the case, i.e. $!(mo \vee we \vee fr)$:

$$M|(mo \vee we \vee fr) : \qquad \boxed{s_1 : mo} \leftarrow - \rightarrow \boxed{s_2 : we} \leftarrow - \rightarrow \boxed{s_3 : fr}$$

The second part of the announcement, (B), is a possibility claim about surprise: 'a surprise should be possible'. Remember the definition of surprise: $surprise := B\varphi \wedge BEFORE(B\neg\varphi)$. In other words, the meaning of (B) is: it should be possible that when the student learns (comes to believe) the exact date of the exam it is the case that previously she had believed that the exam cannot be on that date. First of all, remark that $M|(mo \vee we \vee fr) \not\models K\neg FUTURE\neg surprise$. So, in this model it is not the case that the student knows that whatever sequence of events will follow she will still be surprised. The reason is that if the sequence of events is $!\neg mo; !\neg we; !fr$, the student will not be surprised: $(M|(mo \vee we \vee fr)|\neg mo)|\neg we \not\models surprise$ since $(M|(mo \vee we \vee fr)|\neg mo)|\neg we \models Bfr \wedge \neg BEFORE(B\neg fr)$ (since $M|(mo \vee we \vee fr)|\neg mo \not\models \neg B\neg fr$). Therefore, in order for the teacher's announcement to work in such a way that the student comes to know that she could receive a surprise examination (which would mean that she trusts the teacher with respect to the second part of his announcement), the teacher's announcement has to convey the information that the student is in such a situation such that when she learns when the exam comes it is possible that she will not have expected it. In other words, the second part of the teacher's announcement has to be such that after it, the student comes to know that it is possible she will be surprised: $KFUTUREsurprise$ ('the student knows that there exists a sequence of events after which surprise holds'). So, we can model the second part of the teacher's announcement as a soft upgrade $\otimes\varphi$: $M|(mo \vee we \vee fr) \models [\otimes\varphi]KFUTUREsurprise$ ('after the teacher's announcement the student knows that there exists a sequence of events such that at some point in the future she will be surprised').

Therefore, we can think of the second part of the teacher's announcement as a soft upgrade $\otimes\varphi$ which satisfies the condition that after the product upgrade of $M|(mo \vee we \vee fr)$ with φ the student knows that even if she is not surprised then, she could be surprised at some point in the future (there exists a sequence of events that leads to a state in which *surprise* holds). If the initial model is $M|(mo \vee we \vee fr)$, then there are seven ways in which the student's information could be changed by the act performed by the teacher so that the intended effect to obtain (I omit, for simplicity the transitive closure of the plausibility relations and I define $exam := mo \vee we \vee fr$):

$$(M|exam) \otimes_1 \varphi : \qquad \boxed{s_1 : mo} \vdash - \rightarrow \boxed{s_2 : we} \leftarrow - \rightarrow \boxed{s_3 : fr}$$

$$(M|exam) \otimes_2 \varphi : \qquad \boxed{s_1 : mo} \leftarrow - \dashv \boxed{s_2 : we} \vdash - \rightarrow \boxed{s_3 : fr}$$

$$(M|exam) \otimes_3 \varphi : \qquad \boxed{s_1 : mo} \vdash - \rightarrow \boxed{s_2 : we} \vdash - \rightarrow \boxed{s_3 : fr}$$

$$(M|exam) \otimes_4 \varphi : \qquad \boxed{s_1 : mo} \leftarrow - \rightarrow \boxed{s_2 : we} \vdash - \rightarrow \boxed{s_3 : fr}$$

$$(M|exam) \otimes_5 \varphi : \qquad \boxed{s_1 : mo} \vdash - \rightarrow \boxed{s_2 : we} \leftarrow - \dashv \boxed{s_3 : fr}$$

$(M|exam) \otimes_6 \varphi$:
$$\boxed{s_1 : mo} \leftarrow - - \boxed{s_2 : we} \leftarrow - - \boxed{s_3 : fr}$$

$(M|exam) \otimes_7 \varphi$:
$$\boxed{s_1 : mo} \leftarrow - - \rightarrow \boxed{s_2 : we} \leftarrow - - \boxed{s_3 : fr}$$

It is easy to check that in all of these seven models, surprise could at some point emerge, that is, the student could at some point come to believe that the examination is going to be on a certain day, while before learning that she had believed that it was impossible for the examination to be on that day. For instance take $(M|exam) \otimes_7 \varphi$: after Monday passes with no exam taking place, the student upgrades with $\otimes_7 \varphi$, which leaves her with

$$((M|exam) \otimes_7 \varphi)|\neg mo : \qquad \boxed{s_2 : we} \leftarrow - - \boxed{s_3 : fr}$$

In this new model, the student will believe that the exam is on Wednesday, and she will believe that an exam cannot be on Friday. If Wednesday passes without an exam being given, the student will come to believe that fr, and before coming to believe that fr she had believed that fr cannot be the case. Therefore, surprise is indeed possible if the teacher's action (second part of the announcement) is construed as the upgrade $\otimes_7 \varphi$, and if $!\neg mo; !\neg we$ occurs.

However, *surprise* is by no means necessary yet. There is nothing that prevents the teacher (so far) from giving an examination on a date predictable to the student. For instance, $(M|exam) \otimes_7 \varphi \models K(FUTURE\neg surprise)$, that is 'there is a sequence of events such that if the student interprets the teacher's announcement as $\otimes_7 \varphi$, she will not be surprised by the examination'. That sequence of events is $!\neg mo; !we$. Of course, this is very far from the intuitive meaning of the teacher's announcement and this is why the teacher's announcement has a third part, (C).

The third part of the teacher's announcement is a procedural claim: it tells the student how the teacher will act. The effect of this third part will be not on the number of states the student entertains as possible (as (A) was) or on the plausibility relations of the student (as (B) was), but on the protocol that the student thinks it is running. The third part of the announcement informs the student that the teacher will act in such a way so that the student receives a surprise examination. Thus the worry discussed in the last paragraph dissolves. This procedural claim can be best thought of as a protocol p, 'if the student believes that a certain day cannot be the day of the examination then give the examination on that day'. Let p_i be the (uniform) protocol for model $(M|exam) \otimes_i \varphi$, then the corresponding protocols are:

$$p_1 = \{!mo; !mo!\neg we; !mo!\neg we!\neg fr\},$$
$$p_2 = \{!\neg mo; !\neg mo!we; !\neg mo!we!\neg fr\},$$
$$p_3 = \{!mo; !mo!\neg we; !mo!\neg we!\neg fr\},$$
$$p_3' = \{!\neg mo; !\neg mo!we; !\neg mo!we!\neg fr\},$$
$$p_4 = \{!\neg mo; !\neg mo!we; !\neg mo!we!\neg fr\},$$
$$p_4' = \{!mo; !mo!\neg we; !mo!\neg we!\neg fr\},$$
$$p_5 = \{!mo; !mo!\neg we; !mo!\neg we!\neg fr\},$$
$$p_5' = \{!\neg mo; !\neg mo!\neg we; !\neg mo!\neg we!fr\},$$
$$p_6 = \{!\neg mo; !\neg mo!\neg we; !\neg mo!\neg we!fr\},$$
$$p_7 = \{!\neg mo; !\neg mo!\neg we; !\neg mo!\neg we!fr\}.$$

So, the effect of the last part of the teacher's announcement on the student's information state is (mark the effect of updating an arbitrary model $\mathfrak{M}$ with a protocol, p, as $\mathfrak{M} * p$): $((M|exam) \otimes_i \varphi) * p_i \models \neg FUTURE \neg surprise$. This reads: 'after the teacher's announcement all sequences of events will lead to the student being surprised'. If the necessity of (A) should be obvious to everyone, the necessity of both (B) and (C) might be more obscure. The motivation behind them is the following: if only (C) were the case, then, given that the initial model of the students was M, the teacher would not have been able to give a surprise examination, since it would not have been possible to surprise the student (according to $surprise$) - she initially considers all the states as equally possible (see $surprise$). The problem with only holding (B) was already discussed: the student would not know that the teacher really intends to surprise her. Thus (A), (B) and (C) represent the three distinct parts of the teacher's announcement.

However, there is something unpleasant with saying that the teacher's action can be interpreted as *any* of the seven models above together with their corresponding protocols. First remark that it is obvious that the student knows what surprise means. So, when the teacher announces that the students will receive a surprise examination, he does not only convey that the student has to be in a certain state so that she can indeed be surprised (B), but also that he will act in such a way so that the student will be surprised (C). Therefore, after the teacher's announcement the student also knows that the examination will come only in those days that she consider impossible (since now, after (C), she knows what protocol is running).

The seven models above divide into two categories. On the one hand there are $(M|exam) \otimes_1 \varphi$, $(M|exam) \otimes_2 \varphi$, $(M|exam) \otimes_6 \varphi$, $(M|exam) \otimes_7 \varphi$, and on the other $(M|exam) \otimes_3 \varphi$, $(M|exam) \otimes_4 \varphi$ and $(M|exam) \otimes_5 \varphi$. What distinguishes these two classes is that in the former there is only one world in each model on which if the exam were to come, the student would be surprised, i.e. mo, we, fr and fr, respectively. On the other hand, in the latter class there are two worlds in each model, i.e. mo and we, mo and we, and mo and fr, respectively. The problem with the first set of models is that they have such a structure that the student is able to predict from her beliefs and from the protocol p that the exam will come on the day she considers impossible. Since there is only one such day, then the student is able to predict that at the next round she will learn the exact day of the examination. But this does not seem really surprising: how can you be surprised by φ, if you expect φ to happen?

Therefore, I believe that a small proviso needs to be added to the definition of surprise, namely the proviso that before coming to believe that the exam is on a certain day, the student should not be able to predict that she will learn that. That is, the new and final definition of surprise is: $SURPRISE := B\varphi \wedge BEFORE(B\neg\varphi) \wedge BEFORE(\neg BNEXT(B\varphi))$. It reads 'an agent is surprised by φ if he believes φ, but before he believed that $\neg\varphi$ and, moreover, he was not expecting to come to believe φ'. It is now easy to check that according to this definition of surprise, the models from the first set do not satisfy the intuitive condition that after the announcement the student still has to be surprised, whereas the second set of models do.

$$\forall i \in \{1, 2, 6, 7\} : ((M|exam) \otimes_i \varphi)^* p_i \not\models K\neg FUTURESURPRISE, \text{ whereas,}$$

$$\forall i \in \{3, 4, 5\} : ((M|exam) \otimes_i \varphi)^* p_i \models KFUTURESURPRISE$$

To sum up, the teacher's announcement lets the student know that there is an examination in the following week, it places the student in a state in which she can be surprised and it informs the student that the teacher will surprise her. Therefore, the teacher's announcement contains three components: a public announcement of *exam*; a soft upgrade with sentence φ (which is not important to spell out, since we are more interested in its effect on the student) that places the student in a state in which she can be surprised; and the protocol that the teacher will follow from that moment onwards (i.e. (A)+(B)+(C)).

A few final comments about the way in which surprise was construed in this section. First of all, in principle, the teacher can now give the exam on all days of the week (together, the three ways in which the student can ultimately revise her beliefs after the teacher's announcements, namely $((M|exam) \otimes_3 \varphi) * p_3$, $((M|exam) \otimes_4 \varphi) * p_4$ and $((M|exam) \otimes_5 \varphi) * p_5$, cover all days of the week as possible examination days). However, he cannot give the exam whenever he wants; he has to take into the account whether the student has construed his announcement as $\otimes_3\varphi$, $\otimes_4\varphi$ or $\otimes_5\varphi$. So the teacher has to be aware of how the student understood his announcement. This can easily be taken care of by stipulating that there exists a certain convention between the student and the teacher that together with the teacher's announcement makes it clear to the teacher how the student will revise her beliefs. Or it might be the case that the teacher can rely on past experience or simply that he has some sort of privileged access to the student's reasoning.

Secondly, this solution to the surprise examination paradox makes it a true interactive paradox. The fact that the teacher makes the announcement to the student makes the student possible to be surprised. That is, before the teacher's announcement it was not the case that the student would be surprised (given that she knew an examination was going to take place). The reason is that, before the teacher's announcement, the protocol governing the scenario was the trivial protocol: everything was possible. This means that an unfolding of the scenario with the teacher giving the examination in one of the days the student was foreseeing as examination days could have happened. Moreover, the fact that the student trusts the teacher makes it possible for her to be surprised. If the person making the announcement was not trusted by the student, then she would not revise her beliefs in order to be in a position to be surprised.

Thirdly, and most importantly, it is almost unanimously accepted that the paradoxical feeling that the surprise examination paradox has relies on the teacher's announcement. My analysis of the teacher's announcement as being three distinct actions sheds more light to this idea. I believe that the problematic and *seemingly* contradictory parts of the teacher's announcement are (B) and (C), that is, the possibility claim that the student should be possible to be surprised and the procedural claim saying that the teacher will act so that he will surprise the student. The problematic part is easy to see: if the student is told by what procedure the teacher will try to surprise her, how can she be surprised when the teacher fulfills his own announcement? I agree that this seems contradictory, but I hope the earlier analysis established how this can happen. Namely, there are models (out of seven!) in which the student can be surprised on more than one day, and thus although the student will know that she will learn that the examination is on either of the possible days, she cannot know what exactly she will come to learn.

5 Conclusions

I have used DEL to define surprise in a new way. According to the definition defended in this paper, surprise is a phenomenon related to an act of learning. More precisely, surprise occurs in the moment one learns something that he believed not to be the case before learning it, and moreover, which he did not know that he will learn. Working with this definition, I have defined the teacher's announcement as containing three distinct parts. The first part encodes the hard information that an examination will take place. The second part encodes the soft information that the student has to be in a position to be surprised, while the third part encodes the procedural information that the teacher will only give the examination on a day such that when the student learns which day it is, she will not have believed that it was that day and furthermore, before learning the exact day, the student will not have known that she will learn that the examination is on that day. Furthermore, it easy to see that my solution meets all the conditions of Wright and Sudbury (1977).

There are a few further issues that I have not addressed in this paper, but which deserve attention:

1. The most significant open question which is also of great relevance to my solution is what exactly is the surprise examination paradox. The scenario of surprise examination paradox has been varied in many ways, and all the variations seem to be manifest the same underlying problem. Hence it has been argued that a solution to the surprise examination paradox has to solve all the variations (see Sorensen 1988 and Williamson 2000). However, finding the underlying problem present in all variations is very difficult since no obvious answer seems to be possible. For instance: one variation due to Quine (1953) suggests that the seemingly paradoxical nature of the surprise examination paradox does not depend on the number of classes the student has with the teacher who made the announcement; a variation due to Sorensen (1988) and Williamson (2000) suggests that it does not depend on the assumption that there has to be an examination; another variation due to Sorensen (1988) suggests that it does not depend on the fact that time passes between the teacher's announcement and the examination; and yet another one, also due to Sorensen (1988), suggests that it does not depend on the order in which the days are eliminated; a variation due to Williamson (2000) suggests that it does not depend on the fact that the teacher is making the announcement; a variation due to Ayer (1973) suggests that it does not depend on the fact that the student can doubt the existence of an exam; while a further variation due to Sorensen (1988) suggests that it does not depend on the mental attitude the student has.

My intuition is that a good way of approaching the essence of the surprise examination paradox is through the protocol followed by the teacher and which the student comes to know from his announcement. This would be coherent with a recent attempt of Bovens and Ferreira (2010) to compare probabilistic paradoxes by uncovering their underlying probabilistic protocols. It might be the case that the surprise examination paradox is a good example of showing that the same type of investigation can work in a non-probabilistic setting. However, these intuitions need to be spelled out in a more systematic way and a characterization of the (essence of) surprise examination paradox is still missing.

2. The second question that requires further investigation is what exactly is the connection between the surprise examination paradox and other puzzles such as the prisoner's dilemma, the toxin puzzle, etc.? It has been proposed in the literature that in fact the surprise examination paradox hides in itself a series of other puzzles (see Sorensen 1988 for an overview of the puzzles that seem to be related to the surprise examination paradox). Can a solution to the surprise examination paradox shed any light on these other puzzles? Even if there is no direct way of applying the solution to the surprise examination paradox presented here to those puzzles, it might be relevant to apply the same framework of DEL to solve them. Therefore, interesting new application open for trying to understand the prisoner's dilemma or the toxin puzzle, or vagueness by means of hard, soft and procedural information.

3. Finally, there are a few logical problems surrounding my solution that need to be explored. For example, I claimed that the teacher's announcement contains an announcement about the protocol that the teacher will follow when he will give the examination. This was not at all clear in the formalism. In order to be able to express that we need a logic in which protocols can be announced. An example of such a logic would be Wang (2010). Also, the protocol that seems to be in place has future oriented preconditions. This is not the usual way of defining a protocol. If the preconditions talk about the future, a few complications arise. See Hoshi (2009) for a discussion.

In conclusion, DEL can help us clarify what surprise means and what the teacher's announcement implies in such a way that the surprise examination paradox is solved in an intuitively expected way. Furthermore, such a solution could have interesting philosophical consequences both on the way in which we reason about scenarios of the same sort as the surprise examination paradox, and on the way in which we analyze paradoxes in general.

References

Alchourrón, C., P. Gärdenfors, and D. Makinson (1985). "On the Logic of Theory Change: Partial Meet Contraction and Revision Functions". In: *Journal of Symbolic Logic* 50, pp. 510–530.

Ayer, A. J. (1973). "On a Supposed Antinomy". In: *Mind* 82, pp. 125–126.

Baltag, A. (2009). "SURPRISE!? An Answer to the Hangman, or How to Avoid Unexpected Exams!" Presentation given at the Logic and Interactive Rationality (LIRA) Seminar. Amsterdam.

— (2010). "Dynamic-Doxastic Norms Versus Doxastic-Norm Dynamics". Presentation given at the conference 'Formal Models of Norm Change 2'. Amsterdam.

Baltag, A., L. Moss, and S. Solecki (1998). "The Logic of Public Announcements, Common Knowledge and Private Suspicions". In: *Proceedings of the 7th Conference on Theoretical Aspects of Rationality and Knowledge (TARK '98)*, pp. 43–56.

Baltag, A. and S. Smets (2010). "Multi-Agent Belief Dynamics". Lecture series at NASSLLI 2010.

Binkley, R. (1968). "The Surprise Examination in Modal Logic". In: *Journal of Philosophy* 65, pp. 127–136.

Bovens, L. and J. L. Ferreira (2010). "Monty Hall Drives a Wedge Between Judy Benjamin and the Sleeping Beauty: a Reply to Bovens". In: *Analysis* 70, pp. 473–481.

Gerbrandy, J (2007). "The Surprise Examination in Dynamic Epistemic Logic". In: *Synthese* 155, pp. 21–33.

Gerbrandy, J. (1999). "Bisimulations on Planet Kripke". PhD thesis. Institute for Logic, Language and Computation (ILLC), Universiteit van Amsterdam.

Hoshi, T. (2009). "Epistemic Dynamics and Protocol Information". PhD thesis. Stanford University.

Kaplan, D. and R. Montague (1960). "A Paradox Regained". In: *Notre Dame Journal of Formal Logic* 1, pp. 79–90.

Moore, G. E. (1993). "Moore's Paradox". In: *G.E. Moore: Selected Writings*. Ed. by T. Baldwin. London: Routledge, pp. 207–212.

Pacuit, E. and S. Simon (2011). "Reasoning with Protocols under Imperfect Information". In: *Review of Symbolic Logic* 4, pp. 412–444.

Quine, W. V. O. (1953). "On a So-Called Paradox". In: *Mind* 62, pp. 65–67.

Schwitzgebel, E. (2010). "Belief". In: *Stanford Encyclopedia of Philosophy*.

Shafer, G. (1985). "Conditional Probability". In: *International Statistical Review* 53, pp. 261–277.

Shaw, R. (1958). "The Paradox of the Unexpected Examination". In: *Mind* 67, pp. 382–384.

Sorensen, R. (1988). *Blindspots*. Oxford: Clarendon Press.

— (2009). "Epistemic Paradoxes". In: *Stanford Encyclopedia of Philosophy*.

van Benthem, J. (2006). "The Epistemic Logic of IF Games". In: *Philosophy of Jaakko Hintikka*. Ed. by T. P. of Jaakko Hintikka Randall E. Auxier and L. E. Hahn. Chicago, IL: Open Court Press.

— (2011). *Logical Dynamics of Information and Interaction*. Cambridge: Cambridge University Press.

van Ditmarsch, H. and B. Kooi (2006). "The Secret of my Success". In: *Synthese* 151, pp. 1–32.

Wang, Y. (2010). "Epistemic Modeling and Protocol Dynamics". PhD thesis. Universiteit van Amsterdam.

Williamson, T. (2000). *Knowledge and Its Limits*. Oxford: Oxford University Press.

Wright, C. and A. Sudbury (1977). "The Paradox of the Unexpected Examination". In: *Australasian Journal of Philosophy* 55, pp. 41–58.

CHAPTER 10

Duality and the Equational Theory of Regular Languages

Yann Pequignot

Abstract

On the one hand, the Eilenberg variety theorem establishes a bijective correspondence between varieties of formal languages and varieties of finite monoids.
On the other hand, the Reiterman theorem states that varieties of finite monoids
are exactly the classes of finite monoids definable by profinite equations. Together
these two theorems give a structural insight in the algebraic theory of finite automata. We explain how duality theory can account for the combination of these
two theorems, as was pointed out by Gehrke, Grigorieff, and Pin (2008).

1 Introduction

The theory of formal languages is basically concerned with the description of properties of sequences, which are nothing else than sets of sequences. For any finite non
empty set A of symbols called an alphabet, we define a (formal) *language* as a set of
finite sequences of symbols in A. In this context, finite sequences are called words and
are naturally endowed with the concatenation operation. The set of all words on the
alphabet A forms the free monoid A^* for the concatenation with the empty word as the
neutral element. In order to avoid dealing with such classes as the class of finite sets,
we consider that a finite alphabet A is simply a positive natural number.

The specification of a language requires an unambiguous description of which
words belong to that language. One way to achieve this is by using a machine that
recognises the language. Here we are interested in the simplest model of computation:

the finite state automaton. The languages on an alphabet A recognised by a finite automaton form the Boolean algebra $\text{Rec}(A^*)$ of *recognisable languages*, also called *regular languages* in this case. However, the fact that A^* is a monoid leads to an alternative definition in algebraic terms. We say that a language $L \subseteq A^*$ is *recognised* by a finite monoid M if there exists a surjective monoid morphism $\varphi \colon A^* \to M$ and a subset P of M such that for all word $w \in A^*$

$$w \in L \quad \text{if and only if} \quad \varphi(w) \in P.$$

Then a language L is recognised by a finite state automaton if and only if it is recognised by a finite monoid. Thus, the recognisable languages on A^* are those subsets of A^* for which the membership problem reduces by a monoid morphism to the membership problem for a subset of a finite set.

2 Varieties of Languages and Varieties of Finite Monoids

The connection of the theory of formal languages with logic and especially theoretical computer science motivates the study of classes of languages. The algebraic approach allows us to describe certain classes of languages by means of classes of finite monoids.

Of particular interest in algebra are the classes of finite monoids closed under taking submonoids, quotient monoids, and finite direct products. We call such classes *varieties of finite monoids*.

Any variety of finite monoids $\mathbf{V}$ defines for each finite alphabet A the family of those languages on A recognised by a monoid in $\mathbf{V}$. It can be easily showed directly that the correspondence so defined is a function $\mathcal{V}$ which associate to each finite alphabet A a family of languages $\mathcal{V}(A^*)$ such that

1. for all A, $\mathcal{V}(A^*)$ is a Boolean subalgebra of $\text{Rec}(A^*)$,

2. for all A, $\mathcal{V}(A^*)$ is *closed under quotienting*, that is for all $L \in \mathcal{V}(A^*)$ and for all $u \in A^*$, the languages

$$u^{-1}L = \{w \in A^* \mid uw \in L\}$$
$$Lu^{-1} = \{w \in A^* \mid wu \in L\}$$

   belong to $\mathcal{V}(A^*)$,

3. for all monoid morphisms $\varphi \colon A^* \to B^*$, if $L \in \mathcal{V}(B^*)$ then $\varphi^{-1}(L) \in \mathcal{V}(A^*)$.

Such a function is called a *variety of languages*.

Reciprocally, given a variety of languages $\mathcal{V}$ we associate a variety of finite monoids $\mathbf{V}$ as follows. For each finite alphabet A and each language $L \in \mathcal{V}(A^*)$, we consider the *syntactic monoid M_L* of L defined as the quotient of A^* by the congruence

$$u \sim_L v \quad \text{if and only if} \quad \text{for all } x, y \in A^*(xuy \in L \leftrightarrow xvy \in L).$$

This congruence saturates L and is of finite index so that the corresponding quotient of A^* is a finite monoid recognising L. It is in fact the coarsest congruence which saturates L. We associate to the variety of languages $\mathcal{V}$ the variety of finite monoids $\mathbf{V}$ generated by the syntactic monoids of all languages of $\mathcal{V}$.

The Variety Theorem of Eilenberg then states that

Theorem 2.1 (Eilenberg 1974). *The correspondence between varieties of finite monoids and varieties of languages described above is bijective and order-preserving.*

3 Free Profinite Monoids and the Reiterman Theorem

The Birkhoff theorem (Birkhoff 1935) states that classes of (possibly infinite) monoids are closed under taking submonoids, quotients, and any products if and only if they are definable by equations. Here an equation is a formal equality between elements of a free monoid on a finite set. For example, the equation $xy = yx$ for commutative monoids can be seen as the pair (xy, yx) of words on the alphabet $\{x, y\}$. A monoid M then satisfies the equation (xy, yx) if for all monoid morphism $\varphi \colon \{x, y\}^* \to M$ we have $\varphi(xy) = \varphi(yx)$. Since $\{x, y\}^*$ is the free monoid on $\{x, y\}$ this simply amounts to the condition that for all interpretation $\varphi \colon \{x, y\} \to M$ of x and y in M the induced morphism $\varphi \colon \{x, y\}^* \to M$ equalises the words xy and yx. This is just another way of saying that for all x and y in M, $xy = yx$.

The Reiterman theorem is a counterpart for varieties of finite monoids of the Birkhoff theorem. However, in the case of a variety of finite monoids, the situation is somewhat modified. While classes of finite monoids defined by set of equations of the type just described are indeed varieties of finite monoids, not every variety of finite monoids is defined by a set of such equations. A more general kind of equation is needed. Actually, as it is notably observed in Almeida (2005) and Almeida and Weil (1995), this can be explained by the fact that in general varieties of finite monoids lack free objects. Indeed, in order to have a counterpart to the Birkhoff theorem we have to find objects which relate to finite monoids the same way as the free monoids relate to monoids.

The fact is that the free objects necessary to state the counterpart for finite monoids of Birkhoff theorem are to be taken in the category of *profinite monoids*, whose objects are compact Hausdorff topological monoids which are in addition zero dimensional, i.e. they admit a basis of clopen sets, and whose morphisms are the continuous monoid morphisms.

For the reader initiated to category theory, we can explain this fact as follows. We recall that by the Adjoint Functor Theorem the existence of free objects relates to the completeness of the category, i.e. the existence of all small limits (see for example Lane 1998). Though the (essentially) small category of finite monoids does not enjoy this property, its pro-completion provides an optimal complete category in which the category of finite monoids embeds (Johnstone 1986; Grothendieck and Verdier 1972–1973; Lambek 1966). The pro-completion of the category of finite monoids can be seen as the full subcategory of the category of topological monoids consisting of all projective limits of finite discrete monoids. By general topology and Tychonoff's theorem, any such projective limit is a compact Hausdorff topological monoid which is in addition zero-dimensional. In fact Numakura (1957) proved that any such topological monoid is a projective limit of finite discrete monoids. Hence the pro-completion of the category of finite monoids is (equivalent to) the category of Hausdorff compact zero-dimensional monoids.

For a set A, the *free profinite monoid* on A is defined by the following universal property. It is the unique profinite monoid $\widehat{A^*}$ along with a set function $\iota_A \colon A \to \widehat{A^*}$ such that for all set functions $f \colon A \to M$ into a profinite monoid M there exists a unique continuous monoid morphism $\widehat{f}\colon \widehat{A^*} \to M$ with the following diagram commuting.

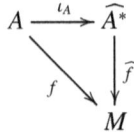

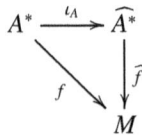

$$A \xrightarrow{\iota_A} \widehat{A^*}$$

with maps f and $\widehat{f}$ to M.

Concretely, it can be defined (see for example Almeida and Weil 1995) as the limit of the following projective system. We consider the family of all finite A-generated discrete monoids, i.e. the maps $\varphi\colon A \to M$ with M a finite monoid such that the image of φ generates M. Then we take maps between a pair $\varphi\colon A \to M$, $\psi\colon A \to N$ of A-generated discrete monoids to be a (continuous) morphism $f\colon M \to N$ such that $f \circ \varphi = \psi$, in particular f must be onto. The limit of this projective system is the map $\iota_A\colon A \to \widehat{A^*}$ defined by the universal property described above.

In fact, by freeness of A^*, A-generated finite monoids $\varphi\colon A \to M$ are in bijective correspondence with monoid quotients $\varphi^*\colon A^* \to M$. The free profinite monoid on A is thus also the projective limit of the finite quotients of A^*. To make this idea precise, let Θ be the set of congruences of finite index on A^* partially ordered with reverse inclusion. This partial order is a directed set since the intersection of two congruences of finite index is again a congruence of finite index. The desired projective system is then the Θ indexed diagram in the category of profinite monoids defined by sending each $\theta \in \Theta$ to the corresponding finite quotient of A^*, namely A^*/θ, and when $\theta \subseteq \theta'$ letting the associated morphism $A^*/\theta \to A^*/\theta'$ be the unique morphism which commutes with the corresponding quotient maps $A^* \to A^*/\theta$ and $A^* \to A^*/\theta'$. The free profinite monoid on A is the limit in the category of profinite monoids of this projective system. Of course, the free profinite monoid on A along with the natural monoid morphism $\iota_A\colon A^* \to \widehat{A^*}$ is also characterised by the fact for all monoid morphisms $f\colon A^* \to M$ into a profinite monoid M there exists a unique continuous monoid morphism $\widehat{f}\colon \widehat{A^*} \to M$ with the following diagram commuting.

$$A^* \xrightarrow{\iota_A} \widehat{A^*}$$

with maps f and $\widehat{f}$ to M.

Remark 3.1. *The free profinite monoid on a finite alphabet A can be obtained in a more pedestrian fashion. Define on A^* a metric d by letting for any words $u, v \in A^*$*

$$d(u, v) = 2^{-\min\{|M| \,\mid\, M \text{ is a finite monoid which separates } u \text{ and } v\}}$$

where $|M|$ denotes the cardinality of M and where a finite monoid M separates two words u and v if there exists a morphism $\varphi\colon A^* \to M$ such that $\varphi(u) \neq \varphi(v)$. It can then be observed that the product on words is uniformly continuous. The free profinite monoid on A is also obtained as the metric completion of (A^*, d) with the monoid operation being the continuous extension of the product on A^*. The equivalence between the two definition can be found in Almeida (1994).

The points in the free profinite monoid on A are called *profinite words* and they turn out to be the right generalisation of terms in order to capture variety of finite monoids by means of equations. The satisfiability of an equation in profinite terms by a monoid is *mutatis mutandis* the same as for word equations. For a pair of profinite words $(x, y) \in \widehat{A^*} \times \widehat{A^*}$ we say that a finite monoid M satisfies the *profinite equation* (x, y) if for any continuous morphism $\varphi \colon \widehat{A^*} \to M$ we have $\varphi(x) = \varphi(y)$. By freeness of $\widehat{A^*}$ this is equivalent to say that for any interpretation of the variables $\varphi \colon A \to M$ the induced continuous morphism $\varphi \colon \widehat{A^*} \to M$ equalises the profinite words x and y. We can now state

Theorem 3.2 (Reiterman 1982). *A class of finite monoids is a variety of finite monoids if and only if it can be defined by profinite equations.*

Note that the crux of the proof of this theorem is the 'only if' part. Furthermore, it should be noticed that defining a variety of profinite monoid may require a set of equations between profinite words taken on unboundedly large finite alphabets.

4 The Eilenberg-Reiterman Theorem

We call the Eilenberg-Reiterman theorem the combination of Eilenberg and Reiterman theorems. By this, we mean that a class of recognisable languages is a variety of languages if and only if it can be defined by profinite equations. Indeed, a variety of languages is, by Eilenberg theorem, recognised by a variety of finite monoids, which in turn is defined by profinite equations by Reiterman theorem. But how does a profinite equation directly define a variety of language?

To see this we let (x, y) be a pair of profinite words on a finite alphabet A and L be a recognisable language on a finite alphabet B. Then the language L is recognised by a finite monoid in the variety of finite monoids defined by the equation (x, y) if, by definition, the syntactic monoid M_L satisfies (x, y). This means that for all continuous morphisms $\varphi \colon \widehat{A^*} \to M_L$ we have $\varphi(x) = \varphi(y)$.

We can in fact be more precise, since each such map $\varphi \colon \widehat{A^*} \to M_L$ arises as the composition of a morphism $\widehat{\psi} \colon \widehat{A^*} \to \widehat{B^*}$ with the quotient map $\widehat{q_L} \colon \widehat{B^*} \to M_L$. Indeed for each $a \in A$, since $\widehat{q_L}$ is surjective, we can choose a $\psi(a)$ in $\widehat{q_L}^{-1}(\varphi(a))$ and let $\widehat{\psi} \colon \widehat{A^*} \to \widehat{B^*}$ be the continuous morphism given by the universal property of $\widehat{A^*}$ for the so defined function $\psi \colon A \to \widehat{B^*}$.

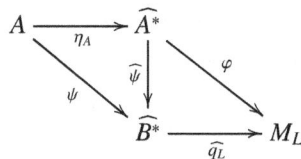

Hence a recognisable language L on a finite alphabet B *satisfies* the profinite equation $(x, y) \in \widehat{A^*} \times \widehat{A^*}$ if for all continuous morphism $\widehat{\psi} \colon \widehat{A^*} \to \widehat{B^*}$ the quotient map $\widehat{q_L} \colon \widehat{B^*} \to M_L$ equalises $\widehat{\psi}(x)$ and $\widehat{\psi}(y)$.

The notion of satisfiablity of a profinite equation by a recognisable language thus defined we can state

Theorem 4.1 (Eilenberg-Reiterman). *A class of recognisable languages is a variety of languages if and only if it can be defined by profinite equations.*

The next sections aims to explain how duality theory can account for this theorem.

5 Stone Duality

Roughly speaking, Stone duality relates intimately Boolean algebras with Stone spaces. The Stone spaces are the Hausdorff compact topological spaces which are zero dimensional, i.e. which admit a basis of clopen sets. We briefly recall what this relation is about (Stone 1936; Johnstone 1986; Givant and Halmos 2009).

For a Boolean algebra B, we denote by X_B the set of all ultrafilters on B endowed by the topology generated by the sets of the form $\{u \in X_B \mid b \in u\}$ with $b \in B$. Equivalently, X_B can be seen as the set of all boolean homomorphisms from B to the two elements Boolean algebra 2, endowed with the topology induced by the product topology on 2^B with 2 discrete. Reciprocally, for a Stone space X we denote by $\mathrm{Clop}(X)$ the set of clopen sets of X endowed with the natural set theoretical Boolean operations. Equivalently, $\mathrm{Clop}(X)$ can be seen as the Boolean subalgebra of the power set 2^X consisting of continuous map $c \colon X \to 2$ with $2 = \{0, 1\}$ carrying the discrete topology. By use of the Boolean Prime Ideal Theorem, these correspondences are reciprocal up to isomorphism, so that $X_{\mathrm{Clop}(X)} = X$ and $\mathrm{Clop}(X_B) = B$. The corresponding isomorphisms are given respectively by

$$\eta \colon B \to \mathrm{Clop}(X_B)$$
$$b \mapsto \{u \in X_B \mid b \in u\}$$

and

$$\epsilon \colon X \to X_{\mathrm{Clop}(X)}$$
$$u \mapsto \{a \in \mathrm{Clop}(X) \mid u \in a\}.$$

Furthermore, this correspondence extends to morphisms in the sense that to each boolean morphism $h \colon B \to B'$ is associated a continuous map $X_h \colon X_{B'} \to X_B$ between the correspondent Stone spaces but in the reverse direction. It is simply defined by $f(u) = u \circ h$ for any $u \in X_{B'}$ seen as Boolean morphism $u \colon B' \to 2$. Reciprocally, to each continuous map $f \colon X' \to X$ between Stone spaces is associated the Boolean morphism consisting in the restriction of the preimage map to clopen sets $\mathrm{Clop}(f) \colon \mathrm{Clop}(X') \to \mathrm{Clop}(X)$ defined for all characteristic function c of a clopen set of X' by $h(c) = f \circ c$. We can thus view Stone duality in terms of the 'double life' of the set 2 being at the same time the discrete space with two elements and the two elements Boolean algebra. In this sense, Stone duality can be depicted by the following diagram.

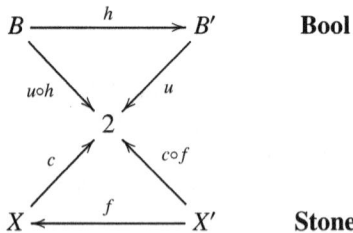

Stone duality states that the category **Bool** of Boolean algebras with Boolean morphisms is equivalent to the category **Stone** of Stone spaces with continuous maps where every arrow is reversed. In the setting of category theory, this amounts to **Bool** $\cong$ **Stone**^{op} and the equivalence is given by the functors $B \mapsto X_B$, $h \mapsto X_h$ and $X \mapsto \mathrm{Clop}(X)$, $f \mapsto \mathrm{Clop}(f)$ together with the natural isomorphisms η and ϵ defined above.

In particular, Stone duality assures that monomorphisms in **Bool** correspond dually to epimorphisms in **Stone**. That is a Boolean morphism $h : A \to B$ is a Boolean embedding if and only if the dual map $X_h : X_B \to X_A$ is a quotient map.

To make this idea precise, let $A \hookrightarrow B$ be an inclusion of Boolean algebras. The dual map $X_B \twoheadrightarrow X_A$ sends each ultrafilter x on B to the ultrafilter $x \cap A$ on A. This quotient map is thus given by identifying any ultrafilters x and y on B for which

$$\text{for all } a \in A \quad (a \in x \leftrightarrow a \in y).$$

Conversely, any quotient map $q : X_B \twoheadrightarrow Y$ in Stone is given by quotienting X_B by the equivalence relation $E = \{(x, y) \in X_B \times X_B \mid q(x) = q(y)\}$. So up to isomorphism, the Boolean algebra dual to Y is the subalgebra A of B consisting of the $a \in B$ such that

$$\text{for all } (x, y) \in E \quad (a \in x \leftrightarrow a \in y).$$

Remark 5.1. *In fact, for X_B the Stone dual to a Boolean algebra B, the polarity $P \subset B \times (X \times X)$ given by*

$$(b, (x, y)) \in P \quad \text{if and only if} \quad b \in x \leftrightarrow b \in y$$

defines an antitone Galois connection between the power sets $\mathcal{P}(B)$ and $\mathcal{P}(X_B \times X_B)$ whose Galois closed sets are respectively the subalgebras of B and the equivalence relations on X_B giving rise to Stone quotients.

6 Stone Duality and Recognition

We first recall that we define recognisable languages of A^* as the Boolean subalgebra of $\mathcal{P}(A^*)$ given by

$$\mathrm{Rec}(A^*) = \left\{ \varphi^{-1}(P) \in \mathcal{P}(A^*) \;\middle|\; \begin{array}{l} \varphi : A^* \to M \text{ is a surjective morphism} \\ \text{onto a finite monoid } M \text{ and } P \in \mathcal{P}(M) \end{array} \right\}$$

$$= \bigcup \left\{ \varphi^{-1}(\mathcal{P}(M)) \;\middle|\; \begin{array}{l} \varphi : A^* \to M \text{ is a surjective morphism} \\ \text{onto a finite monoid } M \end{array} \right\}.$$

The first explicit mention of a relation between duality theory with profinite topologies and automata theory goes back to Pippenger (1997), where we can find the following result.

Theorem 6.1. *The underlying topological space of the free profinite monoid on a finite set A is dual to the Boolean algebra of the recognisable subsets of A^*.*

The following proof is based on the proof given in Gehrke (2009) in a more general setting.

Proof. The free profinite monoid on A is the topological monoid obtained as the projective limit taken in the category of profinite monoids of the finite monoid quotients

M of A^* linked with the continuous morphisms making quotient map commutes. Since forgetful functors preserves limits, the underlying Stone space of $\widehat{A^*}$ is the projective limit of the discrete spaces underlying the finite quotients of A^*. But to each quotient map $\varphi\colon A^* \to M$ onto a finite monoid corresponds the Boolean embedding of $\varphi^{-1}\colon \mathcal{P}(M) \to \mathcal{P}(A^*)$. By definition of $\text{Rec}(A^*)$, this map actually embeds the powerset of M into $\text{Rec}(A^*)$. Hence to the projective limit $\widehat{A^*}$ of the finite quotients M of A^* taken in **Stone** corresponds dually the inductive limit of the subalgebras $\mathcal{P}(M)$ of $\text{Rec}(A^*)$ taken in **Bool**. The latter is just the direct union of the subalgebras $\mathcal{P}(M)$ of $\text{Rec}(A^*)$ which is by definition $\text{Rec}(A^*)$.   □

The profinite words on A can thus be thought of as ultrafilters on the Boolean algebra of recognisable languages. We exploit this fact without mentioning it from now on.

On the basis of this link between profinite monoids and duality, we can already make a first step towards the Eilenberg-Reiterman theorem.

Indeed, by the previous section, and as first observed in Pippenger (1997), the Boolean subalgebras of $\text{Rec}(A^*)$ are the duals of the Stone quotients of $\widehat{A^*}$. Hence given a Boolean algebra of recognisable languages on A, we have dually a Stone quotient $q\colon \widehat{A^*} \twoheadrightarrow X_B$. Setting $E = \{(x, y) \in \widehat{A^*} \times \widehat{A^*} \mid q(x) = q(y)\}$, we have seen that B can be recover as the Boolean algebra of the recognisable languages $L \in \text{Rec}(A^*)$ such that

$$\text{for all } (x, y) \in E \quad (L \in x \leftrightarrow L \in y).$$

Now saying that a recognisable language $L \in \text{Rec}(A^*)$ *satisfies* the profinite equation of the form $x \leftrightarrow y$ for $(x, y) \in \widehat{A^*} \times \widehat{A^*}$ if the condition $L \in x \leftrightarrow L \in y$ is satisfied, we can state the following result of Gehrke, Grigorieff, and Pin (2008)

Theorem 6.2. *A set of recognisable languages on A is a Boolean algebra if and only if it can be defined by a set of equations of the form $x \leftrightarrow y$ for profinite words $x, y \in \widehat{A^*}$.*

The proof that any set of equations of the form $x \leftrightarrow y$ defines a Boolean algebra can be verified in a straightforward manner.

7 Extended Stone Duality and Recognition

In the previous section, we saw that the underlying topological space of the free profinite monoid on a finite set A is the Stone dual of the Boolean algebra of recognisable languages $\text{Rec}(A^*)$. An interesting question here is whether duality can account for the product on $\widehat{A^*}$. By the way duality works, the question must be answered by considering supplementary operations on the Boolean algebra of the recognisable languages. In fact $\text{Rec}(A^*)$ is naturally endowed with supplementary operations. Indeed for any $L, M, N \in \text{Rec}(A^*)$ we first have the lifting of the product on A^*

$$L \cdot M = \{uv \in A^* \mid u \in L \text{ and } v \in M\}$$

and most importantly the *right* and *left residuals* of N by M given by

$$M \backslash N = \{u \in A^* \mid \text{for all } v \in M, vu \in N\},$$
$$N / M = \{u \in A^* \mid \text{for all } v \in M, uv \in N\}.$$

Moreover, these operations are intimately related by the preorder relation of the Boolean algebra $\mathrm{Rec}(A^*)$ which is simply the inclusion preorder. Indeed they satisfy for all $L, M, N \in \mathrm{Rec}(A^*)$ the following property

$$L \cdot M \subseteq N \quad \text{if and only if} \quad L \subseteq N/M \quad \text{if and only if} \quad M \subseteq L \backslash N. \tag{10.1}$$

These operations turn $\mathrm{Rec}(A^*)$ into what is called a *residuated Boolean algebra*, namely a Boolean algebra B together with three binary operations $\cdot$, $\backslash$ and $/$, where $\cdot$ preserves finite joins in each coordinate and where the equation (10.1) is verified for all $L, M, N \in B$.

Remark 7.1. *The recognisable languages of A^* are in fact closed under residuals by any subset of A^*. Indeed if L is recognised by a monoid morphism $\varphi \colon A^* \to M$ and $S \subseteq A^*$, then φ also recognises $S \backslash L$ and L/S since $S \backslash L = \varphi^{-1}(\varphi(S) \backslash \varphi(L))$ and $L/S = \varphi^{-1}(\varphi(L)/\varphi(S))$.*

In general extended Stone duality, the supplementary operations on Boolean algebras are captured by supplementary relations on Stone spaces (Goldblatt 1989). The particular case of extended duality for residuated Boolean algebra we need here is exposed in detail in the master thesis of Dekkers (2008).

Roughly, to each residuated Boolean algebra $(B, \cdot, \backslash, /)$ is associated the Stone dual X_B of B with the ternary relation $R \subseteq X_B \times X_B \times X_B$ defined equivalently by

$$(x, y, z) \in R \quad \text{if and only if} \quad \forall a, b \in B, (a \in x \text{ and } b \in y) \to a \cdot b \in z$$
$$\text{if and only if} \quad \forall a, b \in B, (b \in y \text{ and } a \notin z) \to a/b \notin x \tag{10.2}$$
$$\text{if and only if} \quad \forall a, b \in B, (b \in x \text{ and } a \notin z) \to a \backslash b \notin y. \tag{10.3}$$

Notice that the relation R dual to the supplementary operations on B is not functional from $X \times X$ to X in general. Nevertheless we have the following theorem whose proof can be found in Gehrke (2009) in a more general setting.

Theorem 7.2. *The dual space of $(\mathrm{Rec}(A^*), \cdot, \backslash, /)$ under extended Stone duality is the free profinite monoid $\widehat{A^*}$.*

We go on by characterising the Boolean subalgebras of $\mathrm{Rec}(A^*)$ which are closed under quotienting, as defined in the first part of the present paper, in terms of the residuals operations. By a *Boolean residuation ideal* of $\mathrm{Rec}(A^*)$ we mean a Boolean subalgebra B of $\mathrm{Rec}(A^*)$ such that for all $L \in B$, and for all $K \in \mathrm{Rec}(A^*)$ the languages $K \backslash L$ and L/K belong to B.

Proposition 7.3. *The Boolean algebras of recognisable languages closed by quotients by words are exactly the Boolean residuation ideals of $(\mathrm{Rec}(A^*), \cdot, \backslash, /)$.*

Proof. We first observe that closure under quotients by words amounts to closure under residuals by singletons. In the other direction, we let B be a Boolean algebra of recognisable languages of A^* closed under residuals by singletons. For $L \in B$ and $K \in \mathrm{Rec}(A^*)$, we consider a monoid morphism $\varphi \colon A^* \to M$ onto a finite monoid which recognises L. There exists $P \subseteq M$ such that $\varphi^{-1}(P) = L$ and since M is finite we can choose a finite $K' \subseteq K$ such that $\varphi(K') = \varphi(K)$. We obtain that $K \backslash L = \varphi^{-1}(\varphi(K) \backslash P) = \varphi^{-1}(\varphi(K') \backslash P) = \varphi^{-1}((\bigcap_{v \in K'} \varphi(\{v\})) \backslash P) = \bigcap_{v \in K'} \{v\} \backslash L$. Since K' is finite and B is closed under finite intersection, we have that $K \backslash L \in B$. $\qquad\square$

From the viewpoint of extended duality, the interest for the Boolean residuation ideals I of a residuated Boolean algebra B lies in the fact that the relation defined equivalently by (10.2) and (10.3) on the dual of I behaves nicely with the relation on the dual of B. As a result of which, the Boolean residuation ideals of $\mathrm{Rec}(A^*)$ have a nice dual characterisation as the following next corollary of a result of Dekkers (2008) states.

Theorem 7.4. *The Boolean residuation ideals of* $\mathrm{Rec}(A^*)$ *correspond dually to the profinite monoid quotients of* $\widehat{A^*}$. *That is, a Boolean algebra of recognisable languages B embeds in* $\mathrm{Rec}(A^*)$ *as a Boolean residuation ideal if and only if the extended Stone dual of* $(B, \backslash, /)$ *is a profinite monoid quotient of* $\widehat{A^*}$.

Now we are able to show that Boolean algebras of recognisable languages closed under quotienting are definable by profinite equations. Indeed by Theorem 6.2, we know that any Boolean algebra B of recognisable languages on A^* is defined by a set E of equations of the form $x \leftrightarrow y$ for $x, y \in A^*$. But now if B is in addition closed under quotienting, then it is a residuation ideal of $\mathrm{Rec}(A^*)$ and thus by the previous theorem its extended dual is in fact a profinite quotient of $\widehat{A^*}$. This means that B is also defined by the congruence on $\widehat{A^*}$ generated by E. In other words, if B is closed under quotienting then it can be defined as the set of recognisable languages L on A^* satisfying for all $(x, y) \in E$ the condition

$$\forall z, z' \in \widehat{A^*} \quad zxz' \in L \leftrightarrow zyz' \in L$$

Let us say that a recognisable language L on A^* satisfies the equation of the form $x = y$ for $(x, y) \in \widehat{A^*} \times \widehat{A^*}$ if for all $z, z' \in \widehat{A^*}$ we have $zxz' \in L \leftrightarrow zyz' \in L$. This definition of satisfaction of a profinite equation by a language is consistent with the one given in the third part of the present paper.

We have thus obtained the following result of Gehrke, Grigorieff, and Pin (2008).

Theorem 7.5. *A set of recognisable languages on A^* is a Boolean algebra closed under quotienting if and only if it can be defined by a set of profinite equations of the form $x = y$ for $x, y \in \widehat{A^*}$.*

The previous theorem is a 'local' version of the combination of Eilenberg and Reiterman theorems in the sense that a finite alphabet A is fixed. In order to account for the Eilenberg-Reiterman theorem, we have to consider sets E of equations of the form $x = y$ for profinite words $x, y \in \widehat{A^*}$ on arbitrary large finite alphabet A with the additional condition of being closed under substitution. That is, for any continuous morphism $f \colon \widehat{A^*} \to \widehat{B^*}$ between the free profinite monoids on finite alphabet A and B respectively,

$$\text{for all } (x, y) \in \widehat{A^*} \times \widehat{A^*}, \quad (x, y) \in E \to (f(x), f(y)) \in E.$$

References

Almeida, J. (1994). *Finite Semigroups and Universal Algebra*. Translated from the 1992 Portuguese original and revised by the author. River Edge, NJ: World Scientific.

— (2005). "Profinite Semigroups and Applications". In: *Structural Theory of Automata, Semigroups, and Universal Algebra*, pp. 1–45.

Almeida, J. and P. Weil (1995). "Relatively Free Profinite Monoids: an Introduction and Examples". In: *Semigroups, Formal Languages and Groups* 466, pp. 73–117.

Birkhoff, G. (1935). "On the Structure of Abstract Algebras". In: *Proceedings of the Cambridge Philosophical Society* 31, pp. 433–454.

Dekkers, M. (2008). "Stone Duality: An Application in the Theory of Formal Languages". MA thesis. Radboud University Nijmegen.

Eilenberg, S. (1974). *Automata, Languages, and Machines*. Vol. A and B. Walham, MA: Academic Press.

Gehrke, M. (2009). "Stone Duality and the Recognisable Languages over an Algebra". In: *Algebra and Coalgebra in Computer Science*, pp. 236–250.

Gehrke, M., E. Grigorieff, and J. Pin (2008). "Duality and Equational Theory of Regular Languages". In: *ICALP 2008. Part II*. Ed. by L. Aceto et al. Lecture Notes in Computer Science 5126. Berlin: Springer, pp. 246–257.

Givant, S. R. and P. R. Halmos (2009). *Introduction to Boolean Algebras*. New York, NY: Springer.

Goldblatt, R. (1989). "Varieties of Complex Algebras". In: *Annals of Pure and Applied Logic* 44, pp. 173–242.

Grothendieck, A. and J. L. Verdier (1972–1973). "Préfaisceaux, SGA 4: Théorie des Topos et Cohomologie Etale des Schémas, Tome 1. Théorie des Topos". In: *Lecture Notes in Mathematics* 269.

Johnstone, P. T. (1986). *Stone Spaces*. Cambridge: Cambridge University Press.

Lambek, J. (1966). *Completions of Categories*. Berlin: Springer.

Lane, S. M. (1998). *Categories for the Working Mathematician*. Berlin: Springer.

Numakura, K. (1957). "Theorems on Compact Totally Disconnected Semigroups and Lattices". In: *Proceedings of the American Mathematical Society* 8, pp. 623–626.

Pippenger, N. (1997). "Regular Languages and Stone Duality". In: *Theory of Computing Systems* 30, pp. 121–134.

Reiterman, J. (1982). "The Birkhoff Theorem for Finite Algebras". In: *Algebra Universalis* 14, pp. 1–10.

Stone, M. H. (1936). "The Theory of Representation for Boolean Algebras". In: *Transactions of the American Mathematical Society* 40, pp. 37–111.

CHAPTER 11

Ambiguities for NF

Damien Servais

Abstract

Quine has laid the foundation of his new set theory in his article *New Foundations for Mathematical Logic* (Quine 1937), naturally called New Foundations (NF). He wouldn't have imagined that we were still trying to prove the consistency or the inconsistency of his system for set theory. People have tried several ways to solve this problem, no one succeeded, but a lot of interesting topics have emerged from these researches. In this paper, we will focus on notions stemming from results of Specker (1958) and their consequences in ZFJ, and the general indiscernability results of Specker (1962), leading to the concepts of ambiguities.

1 Introduction

We briefly remind the reader of the basic notions that are fundamental to the study of NF and its subsystems. Let us first recall some concepts from Type Theory (TT), given that NF is built in reference to this system.

TT is a set theory formulated in a multi-sorted language. We have an infinity of variables $x^k, y^k, \dots$ where the superscript is called the *type* of the variable. The atomic formulae are of the form: $x^k = y^k$ and $x^k \in y^{k+1}$. We then build the formulae of the language in a usual way using the classic logical connectors. The formulae in which all of the variables are attached to a type according to the mentioned rules are said to be *stratified*. A formula of the language of first order set theory (e.g. ZF) where we can assign types to the variables, so that it becomes stratified, is called, hereafter, *stratifiable*. And we shall call *stratification* σ an assignment of a type to every variable. TT has only two axioms schemes. Axiom of extensionality: (for any k)

two sets x^k and y^k are equal if and only if they have the same elements; and the scheme of comprehension: $\exists y^{k+1} \forall z^k (z^k \in y^{k+1} \leftrightarrow \varphi^\sigma)$ where y^{k+1} is not free in φ^σ. Lastly, TTU is the theory of types allowing the existence of atoms (Urelemente). Formally, we add an axiom defining the sets: Set stands for any stratified version of $x \in y \rightarrow \mathsf{Set}(y)$ and restrict extensionality to sets, namely *stratified weak extensionality*: any stratified version of $(\forall x S\, et(x))(\forall y S\, et(y))(\forall z (z \in x \leftrightarrow z \in y) \rightarrow (x = y)))$.

NF is a first-order theory with membership and equality as primitive predicates. Again, there are one axiom and a scheme: the axiom of extensionality $\forall x \forall y (\forall z (z \in x \leftrightarrow z \in y) \rightarrow x = y)$ and the scheme of comprehension $\exists y \forall z (z \in y \leftrightarrow \varphi)$ where φ is stratifiable and y is not free in φ. NFU is then the first-order theory whose axioms are Set, weak extensionality, and the scheme of comprehension.

A model of TT is a *typed structure* $\langle M_0, M_1, \ldots; R \rangle$ (with M_i nonempty sets and $R \subset M_i \times M_{i+1}$) satisfying the axioms of TT. As an example, we have the *natural structure*

$$\langle\langle M \rangle\rangle = \langle M, \mathcal{P}(M), \ldots; \in \rangle \models \mathsf{TT}$$

for a nonempty set M. A model of NF will be a first order structure $\langle M, \in_{NF} \rangle$ satisfying the axioms of NF.

As a first interesting link between TT and NF, we remark that $\langle M, R \rangle \models \mathsf{NF}$ if and only if $\langle M, M, \ldots; R \rangle \models \mathsf{TT}$.

The theory NF has been built to get rid of the reduplication of definitions of natural numbers, universe, etc. at each level in TT. Also, it's quickly checked that if φ is a theorem of TT, then φ^+, the stratified sentence obtained by adding 1 to each type in φ, is also a theorem of TT. Even though the opposite is not necessarily true, it's then natural to study the expanded system TT + *Amb* where *Amb* stands for the scheme $\{\varphi \leftrightarrow \varphi^+\}$ (sentences of the form $\varphi \leftrightarrow \varphi^+$ are not, in general, theorems of TT). Specker has proved that NF is consistent if and only if TT + *Amb* is consistent (Specker 1958).

An *isomorphism* between the typed structure $\langle M_0, M_1, \ldots; \in_M \rangle$ and the typed structure $\langle N_0, N_1, \ldots; \in_N \rangle$ is a sequence $\langle f_0, f_1, \ldots \rangle$ such that each f_i is a bijection between M_i and N_i such that, $x^i \in_M y^{i+1} \Leftrightarrow f_i(x^i) \in_N f_{i+1}(y^{i+1})$. An isomorphism between $\mathcal{M} = \langle M_0, M_1, \ldots; \in_M \rangle$ and $\mathcal{M}^+ = \langle M_1, M_2, \ldots; \in_M \rangle$ is called a *(type-)shifting isomorphism*. In that case, the structure $\mathcal{M}$ is then called a *shifting structure*.

Proposition 1.1. *If* $\mathcal{M} = \langle M_0, M_1, \ldots \rangle$ *is a shifting structure, then*

$$\langle M_0, \in_f \rangle \models \varphi(x_1, \ldots, x_k) \text{ iff } \langle M_0, M_1, \ldots \rangle \models \varphi(f^{(n_1)}(x_1), \ldots, f^{(n_k)}(x_k)),$$

where the n_i's *are obtained from a stratification of* φ, $\langle f_0, f_1, \ldots \rangle$ *is the type-shifting isomorphism between* $\mathcal{M}$ *and* $\mathcal{M}^+$, *and* $f^{(i)} = f_i \ldots f_0$.

As a corollary of this proposition, we have Specker's first result (Specker 1958):

Corollary 1.2.

1. $\mathrm{Con}_{\mathsf{NF}}$ *if and only if there exists a shifting model of* TT;

2. $\mathrm{Con}_{\mathsf{NFU}}$ *if and only if there exists a shifting model of* TTU.

2 ZF *with an Automorphism*

Of course, the existence of a shifting model of TT is still unproved. To make things even more interesting for the mathematicians, it's always nice to put these results in the more conventional set theory of ZF. Following this idea, Boffa (1993) and recently Kaye (2010) suggested to study ZFJ, namely ZF with an (external) automorphism J.

We know that two elementary equivalent saturated[1] models having the same cardinality are isomorphic. Also, in ZF + GCH, we know that if M is an infinite model of a theory whose size of language is less than α such that $|M| \leq 2^\alpha$, then there exists a elementary saturated extension of cardinality 2^α (note that we can use recursively saturated models as a refinement of this method; this requires M to be enumerable). Now, we can arrange the automorphism to have different properties. We will focus on two properties, giving birth to nice results on NF and NFU.

It's quickly checked that if there exists a set $a \in M$ such that $\mathcal{P}(a) = J(a)$, then there exists a model of NF. Unfortunately, such an automorphism can't exist: by a classic parity argument on the finite sets $\{\ldots, \ldots X_2, X_1, X_0\}$ with $\mathcal{P}(X_{i+1}) = X_i$, by taking $\mathcal{P}(a) = X$ and $J(a) = X$.

Although an automorphism J moving a set onto its set of subsets is impossible, we can consider the weaker assumption that $|\mathcal{P}(a)| = |J(a)|$. This, in turn, also yields a model of NF. Let f be a bijection between $J(a)$ and $\mathcal{P}(a)$. Consider the set $N = \{ x \in M \mid M \models x \in a \}$ as the domain of the relation $\in_{\mathsf{NF}}$ defined as follows: $x \in_{\mathsf{NF}} y$ if and only if $x \in f(J(y))$. Then, $\langle N, \in_{\mathsf{NF}} \rangle$ is clearly a model of NF.

> **Open Question** : Is the following system (with c a constant added to the language of ZF) consistent?

$$\mathsf{ZFJ} + (\exists \text{ cardinal } c)2^c = Jc.$$

However, one can produce models of ZF with automorphism J having the property that there is a set $a \in M$ such that $\mathcal{P}(a) \subset J(a)$. Therefore, let $N = \{ x \mid M \models x \in a \}$ be the domain. Let $J^{-1}(J(a)\backslash\mathcal{P}(a)) = \text{ATOMS}$ be the set of atoms, and the relation $\in_{\mathsf{NFU}}$ defined by $x \in_{\mathsf{NFU}} y$ if and only if $y \notin \text{ATOMS} \wedge x \in J(y)$, then $\langle N, \in_{\mathsf{NFU}} \rangle$ is a model of NFU. Note that the model here obtained has several properties depending on the choice of the original model M taken for ZF (if you choose a to be finite, well-ordered, ... the universe in $\langle N, \in_{\mathsf{NFU}} \rangle$ will be finite, well-ordered, ...). To sum up, we can say that the following system is consistent:

$$\mathsf{ZFJ} + (\exists \text{ ordinal } \alpha)\alpha < J(\alpha).$$

The idea behind such a construction is expressed in Forster (1995) and slightly developed in Kaye (2010) and consists in reducing the number of atoms, i.e., by trying to make $J(a)$ as close to $\mathcal{P}(a)$ as possible, in order to eventually get rid of them and hopefully get a model of NF. The technique aims at finding a sequence of (external)

---

[1] Forster (1995) suggests to obtain models with automorphism using the Ehrenfeucht-Mostowski theorem.

automorphisms J_i such that $J(a) \geqslant J_1(a) \geqslant \ldots \geqslant \mathcal{P}(a)$, such that $\mathcal{P}(a)$ is the projective limit of $J_i(a)$. Researches have to be made in this direction because it raises more problems than answers for the moment.

3 Ambiguity

The very first notion of ambiguous cardinal (referred to as *typical ambiguity*) has been introduced within the framework of NF after the result of Specker (1962). As an improvement of his previous result (Specker 1958), he showed that the existence of an *ambiguous model* of the TT (a model $\langle M_0, M_1, M_2 \ldots \rangle \models$ TT such that $\langle M_0, M_1, \ldots \rangle \equiv \langle M_1, M_2, \ldots \rangle$) yields the consistency of NF. We restrict our research to a particular case of ambiguous model: the natural structure over a non-empty set M: $\langle\langle M \rangle\rangle = \langle M, \mathcal{P}(M), \ldots \rangle$.

Let M and N be two non-empty sets such that $|M| = |N|$. Then, $\langle\langle M \rangle\rangle \simeq \langle\langle N \rangle\rangle$. We can then define:

Definition 3.1. *A cardinal μ is called an* ambiguous cardinal *if and only if $\langle\langle M \rangle\rangle$ is ambiguous, for any $|M| = \mu$.*

The search of ambiguous cardinals has not been successful. The former idea was to study, as is usually done when working with the problem of the consistency of NF, weaker consistent theories, like NFU, and to enrich them to try to reach a consistent extension corresponding to NF. We have already proved how to find a model of NFU thanks to *weakly ambiguous cardinals* (a weaker notion of ambiguous cardinal), for instance. We clearly see that this method is finer than those usually written in the set theory ZFJ.

There is another way to tackle the problem of the consistency of NF. An idea is to study 'stronger' theories, inconsistent with the existence of a certain kind of ambiguous cardinals, like ZFC, and by weakening them, we would like to find something like the inconsistency of NF. The point is to find interesting definitions of ambiguity in order to make it very fine. Working in ZF and some variant of it makes it possible to define ambiguities with a single formula of ZF.

3.1 SCHEME OF TYPICAL AMBIGUITY

Definition 3.2. *Let the constant a be added to the language of ZF and φ ranging over stratified formulae. We will call the* scheme of natural ambiguity *the set of sentences:*

$$\{\langle\langle a \rangle\rangle \models \varphi \leftrightarrow \langle\langle \mathcal{P}(a) \rangle\rangle \models \varphi\}.$$

The consistency of the natural scheme of ambiguity with ZF yields the consistency of NF.

Seeing that such a consistency proof has remained an open problem for a long time, it's interesting to study other forms of ambiguity. Before that, let us make precise what we mean by different notions of ambiguity. We mean that an *ambiguity* is every notion that yields the natural scheme of ambiguity.

Definition 3.3. *The elements e_1 and e_2 are said to be* indiscernible *if and only if*

$$P(e_1) \leftrightarrow P(e_2)$$

for a specified class of P in a specified system.

We can express ambiguities as particular cases of indiscernabilty.

Definition 3.4. *A first-order formula $P(x)$ is called a* typed property *if there exists a stratified sentence φ such that, for all x,*

$$P(x) \leftrightarrow \langle\langle x \rangle\rangle \models \varphi.$$

Now, the cardinal μ is ambiguous if and only if $\langle\langle x \rangle\rangle \equiv \langle\langle \mathcal{P}(x) \rangle\rangle$, which means that

$$P(\mu) \leftrightarrow P(2^\mu)$$

for all typed properties.

EXAMPLES

1. Let $P(x)$ be the formula expressing the fact that x is not well-ordered. Seeing that Specker has proved that NF $\vdash \neg$AC, by proving that the universe V is not well-ordered, we have a stratified formula φ such that $P(x) \leftrightarrow \langle\langle x \rangle\rangle \models \varphi$.

2. If μ is ambiguous, then so is 2^μ.

3. Let c a cardinal. Let us define $[n]^c$ by induction: $[n]^0 = c$ and $[n+1]^c = 2^{[n]^c}$. We define the *height* of c as: $h(c) = sup\{n | \exists d(c = [n]^d)\}$. Then, for all n, we can see that the formulae $\varphi_n(\mu) \equiv h(\mu) = n$ are typed properties. It turns out that if μ is ambiguous, then $h(\mu) = \infty$.

Obviously, we can define typical ambiguity in stronger ways. Actually, we can see that this notion of ambiguity can be written into a single formula of ZF.

3.2 STRONG AMBIGUITY

An obvious generalization of (typical) ambiguity yields strong ambiguity.

Definition 3.5. *A cardinal μ is* strongly ambiguous *if and only if for all formulae* $P(\mu) \leftrightarrow P(2^\mu)$.

We know that the existence of strongly ambiguous cardinal is inconsistent with ZFC.

Open Question : In the language of ZF with the constant c, is the following system consistent?

$$ZF + \{P(c) \leftrightarrow P(2^c)\}.$$

3.3 N-AMBIGUITY

We will say that a stratified formula φ is *k-stratified* if there are at most k consecutive types in φ. TT_k is the fragment of TT obtained by limiting the formulae to the k first types. We can also definie NF_k as the theory engendered by the axioms of NF that are k-stratifiable.

Then, we can say that a cardinal μ is n-ambiguous if and only if

$$\mu \models \varphi \leftrightarrow 2^\mu \models \varphi$$

for all n-stratified formula φ. We can also say that a formula $P(x)$ is an *n-typed property* if and only if there exists an n-stratified formula S, such that for all x, $P(x) \leftrightarrow \langle\langle x \rangle\rangle \models S$.

Definition 3.6. *A cardinal μ is* n-ambiguous *if and only if it's indiscernible form 2^μ for all n-stratified properties.*

The n-ambiguity has never been really studied except as a consequence of an article by Crabbé (1986). The main result can be written as follows:

Proposition 3.7. *n-ambiguity doesn't imply m-ambiguity for $m > n$.*

Grishin (1969) has proved that $\mathsf{NF} = \mathsf{NF}_4$, entailing that

Proposition 3.8. NF *is consistent if there exists a 3-ambiguous cardinal.*

Moreover, Boffa and Crabbé (1975) proved that every infinite model of TT_3 is ambiguous.

3.4 CUMULATIVE HIERARCHY

As a natural generalization of the n-ambiguity, we would like to study ambiguity for higher type, that means types indexed by ordinals. To achieve such a construction, we have to work with a cumulative structure. So, let us first define a cumulative hierarchy over a non-empty set x_0, seen as a set of *atoms*. So let us consider a non-empty set x and define a cumulative hierarchy as follows (Crabbé 2009):

$$x_0 = USC(x),$$
$$y \in_x z \equiv y \in z \wedge z \notin x_0,$$
$$\mathcal{P}_x(y) = \mathcal{P}(y) \backslash x_0,$$
$$x_{\alpha+1} = x_\alpha \cup \mathcal{P}_x(x_\alpha),$$
$$x_\lambda = \cup_{\alpha<\lambda} x_\alpha.$$

Then we can define the notion of satisfaction for a first order formula A in the first order structure $\langle x_\alpha, \in_x \rangle$ as usual. Note that, if $|x| = |y|$, then $\langle x_\alpha, \in_x \rangle \simeq \langle y_\alpha, \in_y \rangle$. This leads us to define, for a definable ordinal α, $\mu_\alpha \models A$ as

$$\mu_\alpha \models A \Leftrightarrow \forall y(|y| = \mu \rightarrow \langle y_\alpha, \in_y \rangle \models A).$$

The natural notion of ambiguity for a cardinal μ in a cumulative hierarchy, up to an ordinal α, will then be defined as follows:

$$\mu_\alpha \models A \leftrightarrow (2^\mu)_\alpha \models A$$

for all first-order formulae A.

We still don't know whether such notions are inconsistent or not with ZF.

Acknowledgments

This research was supported by the Fonds de la Recherche Scientifique – FNRS.

References

Boffa, M. (1975). "Sets Equipollent to their Power Set in NF". In: *Journal of Symbolic Logic* 40, pp. 149–150.

— (1993). "ZFJ and the Consistency Problem for NF". In: *IPM Proceedings Series (Tehran, Iran)*. Vol. 1, pp. 141–144.

Boffa, M. and M. Crabbé (1975). "Les Théorèmes 3-stratifiés de NF$_3$". In: *Comptes Rendues de l'Académie des Sciences Sér. A-B* 280, pp. 1657–1658.

Crabbé, M. (1986). "Le Schéma d'Ambiguité en Théorie des Types". In: *Bulletin de la Société Mathématique de Belgique (série B)* 38, pp. 46–57.

— (1992). "On NFU". In: *Notre Dame Journal of Formal Logic* 33, pp. 112–119.

— (2009). "Ambiguous Cardinals". In: *Cahiers du Centre de Logique* 16, pp. 89–98.

Forster, T. E. (1995). *Set Theory with a Universal Set (second edition)*. Oxford: Clarendon Press.

Grishin, V. (1969). "Consistency of a Fragment of Quine's NF System". In: *Soviet Mathematics Doklady* 10, pp. 1387–1390.

Jensen, R. (1969). "On the Consistency of a Slight(?) Modification of Quine's NF". In: *Synthese* 19, pp. 250–263.

Kaye, R. (2010). "Automorphisms and Constructions of Models of Set Theory". In: *Cahiers du Centre de Logique* 17, pp. 73–88.

Quine, W. V. O. (1937). "New Foundations for Mathematical Logic". In: *Mathematical Monthly* 44, pp. 70–80.

Servais, D. (2009). "A Note on Ambiguous Cardinals". In: *Logic, Philosophy and History of Science in Belgium. Proceedings of the Young Researchers Days 2008*. Ed. by E. Weber et al. Brussel: Koninklijke Vlaamse Academie van België, pp. 55–60.

Specker, E. P. (1962). "Typical ambiguity". In: *Logic, Methodology and Philosophy of Science*. Ed. by E. Nagel. Stanford, CA: Stanford University Press, pp. 116–123.

Specker, E. (1958). "Dualität". In: *Dialectica* 12, pp. 451–465.

CHAPTER 12

Axioms for Non-Archimedean Probability (NAP)

Sylvia Wenmackers, Vieri Benci, and Leon Horsten

Abstract

In this contribution, we focus on probabilistic problems with a denumerably or non-denumerably infinite number of possible outcomes. Kolmogorov (1933) provided an axiomatic basis for probability theory, presented as a part of measure theory, which is a branch of standard analysis or calculus. Since standard analysis does not allow for non-Archimedean quantities (*i.e.* infinitesimals), we may call Kolmogorov's approach 'Archimedean probability theory'. We show that allowing non-Archimedean probability values may have considerable epistemological advantages in the infinite case. The current paper focuses on the motivation for our new axiomatization.

1 Introduction

Kolmogorov (1933) presents probability theory as a part of measure theory, which is a branch of standard analysis or calculus. Since standard analysis does not allow for non-Archimedean quantities (*i.e.* infinitesimals), we may call Kolmogorov's approach an 'Archimedean probability theory'. We show that allowing non-Archimedean probability values may have considerable epistemological advantages in the infinite case.

In Section 2, we review the axiomatization of probability theory by Kolmogorov (1933) and problems associated with this approach. Kolmogorov (1933) deals with infinite outcome spaces primarily by the Axiom of Continuity. In combination with finite additivity, this axiom leads to σ- or countable additivity in cases where the event space is a σ-algebra. However, the Axiom of Continuity and the property of countable additivity (CA) give rise to a number of epistemological issues, such as:

(1) An event with probability 1 is not necessarily certain to occur.

(2) An event with probability 0 is not necessarily impossible to occur.

(3) It turns out to be impossible to model certain simple problems, such as a fair lottery on the natural numbers.

In Section 3, we conclude that in order to solve all of the interrelated issues, it does not suffice to make a minor adaptation of one of the axioms. Instead, we propose a new system of axioms in Section 4. Our solution generalizes the analysis made in Wenmackers and Horsten (2011) to probabilistic problems not only on other countably infinite sample spaces (*e.g.* a fair lottery on $\mathbb{Q}$), but on sample spaces of larger cardinalities as well (*e.g.* a fair lottery on $\mathbb{R}$).

The current paper focuses on the motivation for our new axiomatization. The consistency proof, the application to examples, and the refutation of philosophical arguments against infinitesimal probabilities are all left for a future article (Benci, Horsten, and Wenmackers 2011).

2 Kolmogorov's Axiomatization

In this section, first we review Kolmogorov's axioms for probability theory. Then, we give an overview of known problems with Kolmogorov's axiomatization of probability theory. Kolmogorov's probability theory works fine as a mathematical theory, but the *direct* interpretation of its language leads to counterintuitive results. Also the fact that some seemingly simple situations cannot be described within Kolmogorov's system needs some epistemological considerations. This global analysis will lead us to the solution strategy of the next section: instead of fighting the symptoms one-by-one, we will carefully select a new set of axioms. These axioms are stated in the final section.

2.1 KOLMOGOROV'S AXIOMS

Here, we present axioms that are equivalent to the original axiomatization of Kolmogorov (1933). In particular, K4 is not Kolmogorov's Continuity Axiom, but rather (an equivalent formulation of) the property of Countable Additivity, which follows from the Continuity Axiom and Finite Additivity.

(K0) **Domain and range.** The events are the elements of a σ-algebra $\mathfrak{A} \subseteq \mathcal{P}(\Omega)$ and the probability function takes the following form:

$$P : \mathfrak{A} \to \mathbb{R}$$

(K1) **Positivity.** $\forall A \in \mathfrak{A}$,
$$P(A) \geq 0.$$

(K2) **Normalization.**
$$P(\Omega) = 1.$$

(K3) **Finite additivity.** If A and B are events,

$$A \cap B = \emptyset \implies P(A \cup B) = P(A) + P(B).$$

(K4) **Countable additivity.** *Let*

$$A = \bigcup_{j=0}^{\infty} A_j$$

with $(\forall j \in \mathbb{N}) A_j \subseteq A_{j+1}$; *then*

$$P(A) = \sup_{j \in \mathbb{N}} P\left(A_j\right).$$

Furthermore, we may split the axiom (K0) into two further parts:

(K0A) **Domain.** The domain of P is a σ-algebra $\mathfrak{A} \subseteq \mathcal{P}(\Omega)$.

(K0B) **Range.** The range of P is (a subset of) $\mathbb{R}$.

From the combination of axioms (K0B) and (K1), we see that the range of P is $\mathbb{R}^+$. This set provides a structure that allows for addition and multiplication of probability values. When we also take into account the Normalization axiom (K2), we obtain that:

$$P : \mathfrak{A} \to [0, 1]_{\mathbb{R}}$$

where $[0, 1]_{\mathbb{R}}$ is the unit interval in $\mathbb{R}$.

2.2 PROBLEMS WITH KOLMOGOROV'S AXIOMATIZATION

Non-Measurable Sets. The elements of the domain of the probability function, $\mathfrak{A}$, are called events. A peculiarity of axiom (K0A) is that it allows $\mathfrak{A} \neq \mathcal{P}(\Omega)$. In fact, it is well-known that there are (probability) measures (such as the Lebesgue measure on $[0, 1]$) which cannot be defined for all the sets in $\mathcal{P}(\Omega)$. Thus, there are sets in $\mathcal{P}(\Omega)$ which are not events, which means that their probability value is undefined. This may even occur for sets that are the union of elementary events in $\mathfrak{A}$.

Problems with the Interpretation of $P = 0$ **and** $P = 1$ **Events.** It seems natural to interpret events that have zero probability as impossible events. This idea lies at the basis of the concept of 'Regularity', which can be stated formally as follows:

Regularity. For any event A:

$$P(A) = 0 \Leftrightarrow A = \emptyset. \tag{12.1}$$

Observe that axiom (K1) does allow for events other than the empty set to have probability zero. In particular, when Ω is infinite, it may occur that possible events have probability zero. Thus, whereas Kolmogorov's axioms do guarantee the implication from right to left, they fail to secure the implication in the opposite direction. Hence, Kolmogorov's approach violates Regularity.

In particular, there are situations such that Ω is infinite and *all* the elementary events have probability zero. Then it seems like we have:

$$P(E_j) = 0, \quad j \in J \tag{12.2}$$

and

$$P\left(\bigcup_{j \in J} E_j\right) = 1. \tag{12.3}$$

This situation is very common when J is not denumerable. It looks as if (12.2) states that each event E_j is impossible, but (12.3) states that one of them will occur certainly. In this case, not only Regularity is violated, but also CA, which is axiom (K4) of Kolmogorov's system, fails. In other words, this situation cannot be described in the system at all.

The interpretation of $P = 1$ events turns out to be just as problematic as the interpretation of $P = 0$ events: axiom (K2) does not forbid that a set $A \subsetneq \Omega$ has probability one, and indeed, when Ω is infinite, the occurrence of an event with probability one is not guaranteed to be certain. Therefore, Easwaran (2010) proposed to let 'Regularity' also refer to the 'twin' of (12.1), which says that unit probability should be reserved for the certain event. This can be formalized as follows:

Regularity'. For any event A:

$$P(A) = 1 \Leftrightarrow A = \Omega. \tag{12.4}$$

Again, Kolmogorov's axioms only ensure the implication in the right-to-left direction.

For those who regard Regularity (and its twin) merely as a convenience, there is an obvious solution at hand: interpret probability 0 as 'very unlikely' (rather than simply as 'impossible'), and interpret probability 1 as 'almost certain' (instead of 'absolutely certain'). Yet, there is a philosophical prize to be paid to avoid these contradictions: the correspondence between mathematical formulas and reality is now quite vague—just how probable is 'very likely' or 'almost certain'?—and far from intuition. Moreover, this solution is not acceptable for authors who regard Regularity as a norm of rationality, such as Skyrms (1980) and Lewis (1986). These authors have suggested that Regularity can be restored by allowing infinitesimals in the range of the probability function.

Problems with Conditional Probability. The fact that in Kolmogorov's theory possible events may have probability zero also leads to problems with conditional probability, as defined by (D2). Popper (1938) developed his own basis for probability theory (see also the new appendix *iv in the reprinted version of Popper 1959), consisting of six axioms, in which conditional probability—-denoted as $p(a, b)$—is fundamental. Within his system, conditionalizing on $P = 0$-events does not pose a particular difficulty. However, Popper limits his system to situations with at most denumerably many elements. Thus, his approach does not pose a general alternative to conditionalization on $P = 0$-events.

Fair Lotteries. Considering fair lotteries, we discover two thing: (i) that the range $[0, 1]_\mathbb{R}$ is neither necessary nor sufficient, and (ii) that some very simple problems cannot be dealt with in Kolmogorov's formalism.

Regarding (i), the range of P in case of fair lotteries as an example, we can make the following observations:

- For a fair finite lottery, the unit interval of $\mathbb{R}$ is not necessary as the range of the probability function: the unit interval of $\mathbb{Q}$ suffices.

- In the case of a fair lottery on $\mathbb{N}$, $[0, 1]_\mathbb{R}$ is not sufficient as the range: it violates our intuition that the probability of any set of tickets can be obtained by adding the probabilities of all individual tickets.

In order to illustrate (ii), let us focus on the fair lottery on $\mathbb{N}$ (Finetti 1974). In this case, the sample space is $\Omega = \mathbb{N}$ and we expect the event space (domain of P) $\mathfrak{D}$ to contain all the singletons of $\mathbb{N}$—otherwise there would be 'tickets' (individual, natural numbers) whose probability is undefined, which would be strange for we know they are equal in a fair lottery. Also, we expect to be able to assign a probability to any possible combination of tickets. This assumption implies that the range $\mathcal{E} = \mathcal{P}(X)$ (though not necessarily that $\mathfrak{D} = \mathcal{P}(\Omega)$). Moreover, we expect to be able to calculate the probability of an arbitrary event by some sort of summing over the individual tickets.

The critical inspection of this example leads us to two important observations:

A If we want to have a probability theory which describes a fair lottery on $\mathbb{N}$, assigns a probability to all singletons of $\mathbb{N}$, and follows a generalized additivity rule as well as the Normalization Axiom, the probability has to be non-zero but smaller than any finite, strictly positive real number. Hence, the range of P has to include infinitesimals. In other words, the range of P has to be a subset of a non-Archimedean field. Therefore, it cannot be $\mathbb{R}^+$, but it could be a non-standard set such as $\mathbb{Q}^{*,+}$ or $\mathbb{R}^{*,+}$, which are known from non-standard analysis (NSA).

B Our intuitions regarding infinite concepts are fed by our experience with their finite counterparts. So, if we need to extrapolate the intuitions concerning finite lotteries to infinite ones we need to introduce a sort of limit-operation which transforms 'extrapolations' into 'limits'. Clearly, this operation cannot be the limit of classical analysis. Since the latter is used in Kolmogorov's Finite Additivity, axiom (K4) is suspect.

Motivated by the case study of a fair infinite lottery, at this point we know which elements in Kolmogorov's classical axiomatization we do not accept: the use of $[0, 1]_\mathbb{R}$ as the range of the probability function and the application of classical limits in the Continuity Axiom. However, we have not offered an alternative to his approach, yet: this is what we present in the next section.

3 Solution Strategy

In order to solve one particular element of the problems mentioned, it may suffice to tinker with a single axiom. For instance, in order to be able to describe a fair lottery on the natural numbers, one seems to have the following options:

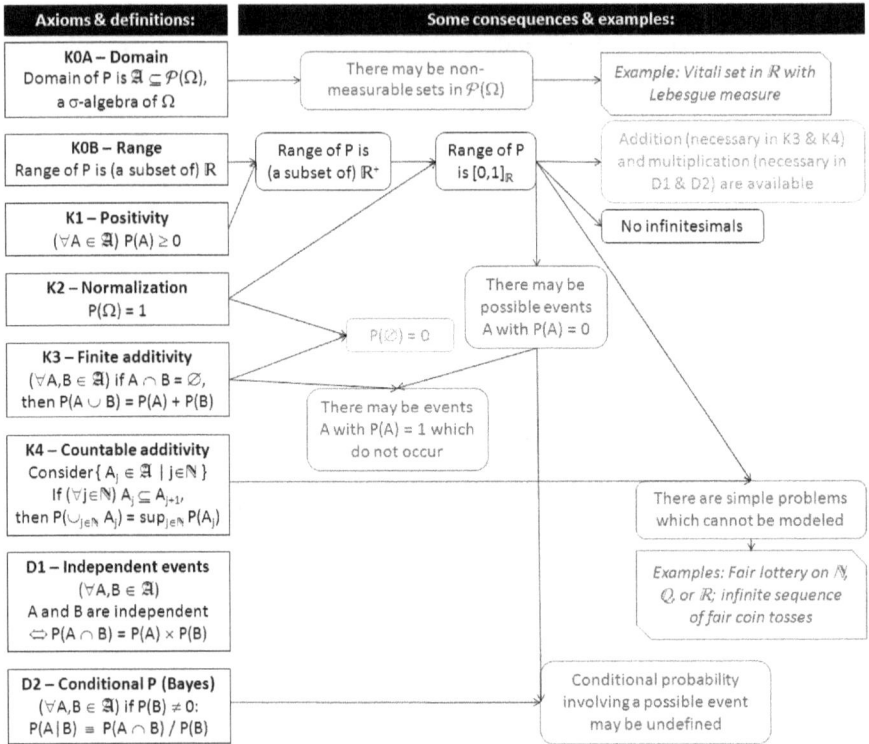

Figure 12.1: Schematic representation of Kolmogorov's axioms and definitions on the left hand side and their consequences (with some examples) on the right hand side. Positive features are indicated in green, negative properties in red.

- Drop the axiom of Normalization (K2). This solution was explored by Rényi (1955).

- Drop the axiom of CA (K4). This solution was advocated by Finetti (1974).

- Change (K0B), replacing $\mathbb{R}$ by $\mathbb{R}^*$, in order to allow infinitesimals in the range of the probability function. This was suggested by Skyrms (1980), Lewis (1986) and Jackson, Priest, and Elga (2004).[1]

Note that none of the above solutions generalizes to allow for the description of probabilistic problems on sample spaces of larger cardinalities as well (*e.g.* a fair lottery on $\mathbb{R}$).

Likewise, if another problem is our major concern, we can find some axiom to adjust. However, from Figure 12.1, we see that *all* of Kolmogorov's axioms are involved in some of the epistemological problems mentioned. In particular, this chart shows that, for cases dealing with infinite sample spaces, the main problem comes from the

---

[1]On closer inspection, this approach also implies that (K4) has to be dropped, as observed in Wenmackers and Horsten (2011).

combination of the choice of the range of P with the axiom of CA (K4). Therefore, if we want to cure all symptoms simultaneously, we should formulate an entire, new set of axioms to base our theory of probability on. Here, we list our main requirements for the new axioms:

- Whereas (K0A) allows for a domain which is strictly smaller than the full power set, it would be convenient if every set in $\mathcal{P}(\Omega)$ represents an event. For finite problems, the maximum number of different values taken by the function P depends on the problem (in particular, the size of the sample space Ω). Therefore, instead of fixing the range in advance, we will allow the range to depend on Ω for the finite as well as the infinite case.

- We intend to replace (K1) by a formalization of Regularity. We should replace (K2) in a similar way, as to ascertain that also the interpretation of $P = 1$ events is intuitively clear.

- The axiom of Finite Additivity (K3) is involved in one problematic issue, but does not seem the main culprit. So far, it seems as though we can keep this principle intact.

- Our replacement for (K4) will be the most drastic removal from Kolmogorov's approach. Because (K4) implies the use of classical limits, it is incompatible with our planned introduction of non-standard numbers in the range of P (*cf.* Wenmackers and Horsten 2011). Our new axiom will imply a limit operation of a different kind. In the case of a denumerably infinite sample space, this limit turns out to be the α-limit, as defined by Benci and Di Nasso (2003) in the context of Alpha-Theory—an axiomatic approach to NSA. In the case of a non-denumerably infinite sample space, a generalization of the ideas underlying the α-limit will have to be derived.

We will translate the above considerations into axioms in the next section.

4 Axioms for Non-Archimedean Probability (NAP)

Since the range of the probability function may contain infinitesimals in our approach, we call it Non-Archimedean Probability (NAP).

These are the Axioms of NAP, where Ω is a set called the sample space:

(NAP0) The probability function P has as its domain the full powerset of Ω (event space = $\mathcal{P}(\Omega)$) and as its range the unit interval of a suitable, ordered field F (range = $[0, 1]_F$).

(NAP1) $P(A) = 0 \Leftrightarrow A = \emptyset$.

(NAP2) $P(A) = 1 \Leftrightarrow A = \Omega$.

(NAP3) For all events $A, B \in \mathcal{P}(\Omega)$ such that $A \cap B = \emptyset$:

$$P(A \cup B) = P(A) + P(B).$$

(NAP4) There exists a directed set[2] $\langle \Lambda, \subseteq \rangle$ with $\Lambda \subseteq \mathcal{P}_{fin}(\Omega)$ such that:[3]

$$\forall E \in \mathcal{P}_{fin}(\Omega), \exists \lambda \in \Lambda : E \subset \lambda.[4]$$

When the sample space Ω is finite, the NAP-axioms have no advantage over Kolmogorov's axioms. This should not be surprising, as they were intended to overcome difficulties that arise only when considering problems on infinite sample spaces. We see that the NAP-axioms indeed fulfill the desiderata of the previous subsection:

- The axiom (NAP0) ensures that every set in $\mathcal{P}(\Omega)$ represents an event. Whereas Kolmogorov fixes the range of all probability functions (as $[0, 1]_{\mathbb{R}}$), in our case, the range depends on the problem (in particular, the sample space Ω).

- The pair (NAP1) & (NAP2) makes it possible to interpret probability 0 and 1 events safely as impossible and necessary, respectively: they formalize Regularity and Regularity', respectively.

- The axiom (NAP3) is exactly the same as Kolmogorov's addition-rule for events (finite additivity).

- The axiom (NAP4) implies the use of a generalized limit (direct limit), just like the Continuity Axiom implies the use of classical limits.

Collectively, the axioms may force the range $[0, 1]_F$ to be non-Archimedean. In particular, when the sample space Ω is countably infinite and the odds are fair, or when Ω is uncountably infinite, (NAP1) forces F to be a non-Archimedean field.

In the case of a denumerably infinite sample space Ω, the limit implied by (NAP4) can be the axiomatically defined α-limit. A non-standard model for this concept can be obtained in terms of maximal ideals or free ultrafilters; the use of the directed set $\langle \Lambda, \subseteq \rangle$ has the advantage that unlike the other two options, Λ can be stated explicitly and can be chosen such as to model 'the physics' of the problem. The meaning of the latter statement will become clear in the next subsection, where we will apply the axioms to relevant examples.

There is another analogy between Kolmogorov's axiom (K4) and the NAP-axiom (NAP4). Axiom (K4) allows to extend the probability to a σ-algebra once you have defined the probability on a smaller family of sets. For example, if you want to define a uniform probability on $[0, 1]_{\mathbb{R}}$, you start by defining the probability on the family of semi-open intervals $(\forall a > b \in [0, 1]_{\mathbb{R}})P([a, b)) = b - a$. Then you extend the set of semi-open interval to the σ-algebra generated by them. The Axiom (NAP4) has the same role: if you want to define a probability on $\mathcal{P}(\Omega)$, you start by defining the probability on a suitable family of finite sets first, and then you extend it to the whole power set.

The current theory can be applied to a wide range of cases: countable lotteries (fair lottery on $\mathbb{N}$ or $\mathbb{Q}$), uncountable ones (fair lottery on $\mathbb{R}$), and infinite sequences of tosses with a fair coin. This means that we can express the probability of a particular

---

[2] $\langle \Lambda, \subseteq \rangle$ is a directed set if and only if Λ is a non-empty set and $\subseteq$ is a preorder such that every pair of elements of Λ has an upper bound: $(\forall A, B \in \Lambda)\exists C \in \Lambda : A \cup B \subseteq C$.
[3] $\mathcal{P}_{fin}(\Omega)$ denotes the family of *finite* subsets of Ω.
[4] This condition implies that $\bigcup_{\lambda \in \Lambda} \lambda = \Omega$.

outcome of an infinite sequence of coin tosses as an infinitesimal, *pace* Williamson (2007). The development of these examples, however, goes beyond the scope of this contribution and will be presented in Benci, Horsten, and Wenmackers (2011).

References

Benci, V. and M. Di Nasso (2003). "Alpha-Theory: an Elementary Axiomatic for Non-standard Analysis". In: *Expositiones Mathematicae* 21, pp. 355–386.

Benci, V., L. Horsten, and S. Wenmackers (2011). "Non-Archimedean Probability". In preparation. URL: http://arxiv.org/abs/1106.1524.

Easwaran, K. (2010). "Regularity and Infinitesimal Credences". Unpublished manuscript.

Finetti, B. de (1974). *Theory of Probability*. Translated by: A. Machí and A. Smith. Hoboken, NJ: Wiley.

Jackson, F., G. Priest, and A. Elga (2004). "Infinitesimal Chances and the Laws of Nature". In: *Australasian Journal of Philosophy* 82, pp. 67–76.

Kolmogorov, A. N. (1933). *Grundbegriffe der Wahrscheinlichkeitrechnung*. Ergebnisse der Mathematik. Translated by N. Morrison, *Foundations of Probability*. London: Chelsea Publishing Company, 1956 (second edition).

Lewis, D. K. (1986). *Philosophical Papers. Volume II*. Oxford: Oxford University Press.

Popper, K. R. (1938). "A Set of Independent Axioms for Probability". In: *Mind* 47, pp. 275–277.

— (1959). *The Logic of Scientific Discovery*. London: Hutchinson.

Rényi, A. (1955). "On a New Axiomatic Theory of Probability". In: *Acta Mathematica Hungarica* 6, pp. 285–335.

Skyrms, B. (1980). *Causal Necessity*. New Haven, CT: Yale University Press.

Wenmackers, S. and L. Horsten (2011). "Fair Infinite Lotteries". In: *Synthese (forthcoming)*. DOI: 10.1007/s11229-010-9836-x.

Williamson, T. (2007). "How Probable is an Infinite Sequence of Heads?" In: *Analysis* 67, pp. 173–180.

CHAPTER 13

Modifying Kremer's Modified Gupta-Belnap Desideratum

Stefan Wintein

Abstract

In a recent paper, Philip Kremer proposes a formal and theory-relative desideratum for theories of truth that is spelled out in terms of the notion of 'no vicious
reference'. Kremer's *Modified Gupta-Belnap Desideratum* (**MGBD**) reads as follows: if theory of truth **T** dictates that there is no vicious reference in ground
model M, then **T** should dictate that truth behaves like a classical concept in M.
In this paper, we suggest an alternative desideratum (**AD**): if theory of truth **T** dictates that there is no vicious reference in ground model M, then **T** should dictate
that all T-sentences are (strongly) assertible in M. We illustrate that **MGBD** and
AD are not equivalent by means of a *Generalized Strong Kleene theory of truth*
and we argue that **AD** is preferable over **MGBD** as a desideratum for theories of
truth.

1 Introduction

In the paper *How Truth Behaves When There's No Vicious Reference*, Kremer (2010) is
concerned with the behavior of truth under circumstances in which there is *no vicious
reference*. Roughly, vicious reference is that type of reference that forces truth—or
the truth predicate—to behave in a non-standard manner. The reference involved in a
Liar sentence certainly is vicious, while the reference involved in (13.1) certainly is
not.

$$(13.1) \text{ consists of 6 words.} \tag{13.1}$$

Kremer argues that our intuitions concerning which sentences exhibit vicious reference and which do not, are (partly) determined by our intuitions concerning which theory of truth is correct. According to Kremer, this influence of truth-theoretic intuitions on vicious reference intuitions is inevitable.

> The most general formal articulation of non-vicious reference, we suggest, will be theory-relative.   Kremer 2010, p. 357

Kremer provides a formal, theory-relative articulation of non-vicious reference and he uses this notion to spell out a formal, theory-relative desideratum for theories of truth. Intuitively, the desideratum, called the *Modified Gupta-Belnap Desideratum* (**MGBD**), says that if there is no vicious reference according to a theory of truth **T**, then, according to **T**, truth should behave like a classical concept. Formally:

MGBD If **T** *dictates that there is no vicious reference in ground model M*, then **T** *dictates that truth behaves like a classical concept in ground model M.*

Kremer compares thirteen theories of truth (ten fixed point theories, three revision theories) in terms of **MGBD**. With respect to the *rationale* of **MGBD**, Kremer cites[1] Gupta (1982), who says that:

> For models *M* belonging to a certain class—a class that we have not formally defined but which in intuitive terms contains models that permit only benign kinds of self-reference—the theory should entail that all Tarski biconditionals are assertible in the model *M*.   Gupta 1982, p. 19

Thus, the proposed rationale for **MGBD** is that it is a theory-relative formalization of Gupta's intuitively stated, theory-neutral desideratum—note, Gupta speaks of an *adequacy condition*—for theories of truth. In this paper, we propose an **A**lternative formal and theory-relative translation of Gupta's intuitive **D**esideratum.

AD If **T** *dictates that there is no vicious reference in M*, then **T** *dictates that all the T-sentences*[2] *are strongly assertible in M*, where a sentence σ is strongly assertible just in case it is assertible and $\neg\sigma$ is not.

Although any theory which violates **AD** violates **MGBD**, we will see that there are theories of truth which violate **MGBD** and satisfy **AD**. When restricted to the thirteen theories of truth considered by Kremer however, **AD** and **MGBD** are equivalent. The reason of this is that all thirteen theories recognize a *single* semantic value which is allotted to all strongly assertible sentences. This semantic value is, per definition, the same value that is allotted to all *classical* strongly assertible sentences, such as 'snow is white'. Accordingly, with **T** one of theories considered by Kremer, **T** dictates that truth behaves as a classical concept in *M* just in case **T** dictates that all the *T*-sentences are strongly assertible in *M*.

Wintein (2011) defined the notion of a *Generalized Strong Kleene theory of truth,*

---

[1]On page 348 of Kremer (2010).

[2]A *T*-sentence, or, in Gupta's words, a Tarski biconditional, is a sentence of form $T(\overline{\sigma}) \leftrightarrow \sigma$, with $\overline{\sigma}$ a closed term which denotes σ.

or GSK theory. The distinction between a (three- or four-valued) Strong Kleene theory of truth and a GSK theory, is that the latter recognizes more than one sense in which a sentence can be strongly assertible. Formally, the semantics of a GSK theory differs from the semantics of a Strong Kleene theory only with respect to negation. Our running example of a GSK theory will be $\mathcal{K}^5$, which has a linear five-valued (generalized) Strong Kleene semantics with respect to the lattice:

$$\leq_5 := \mathbf{d}_g \leq \mathbf{d}_i \leq \mathbf{e} \leq \mathbf{a}_i \leq \mathbf{a}_g.$$

Conjunction and disjunction act as meet and join in $\leq_5$, and universal and existential quantification act as generalized conjunction and disjunction. Negation acts as the identity operation on $\mathbf{e}$ but also, it interchanges $\mathbf{a}_x$ for $\mathbf{d}_x$, where $x \in \{g, i\}$ indicates the *assertoric sense* under consideration: grounded or intrinsic. The notions of groundedness and intrinsicness reflect that $\mathcal{K}^5$ is defined in terms of Kripke's *Strong Kleene minimal fixed point theory* ($\mathcal{K}$) and his *Strong Kleene maximal intrinsic fixed point theory* ($\mathcal{K}^+$).

According to $\mathcal{K}^5$, a sentence is *strongly assertible* just in case its value is contained in $\{\mathbf{a}_g, \mathbf{a}_i\}$, while a sentence is *classical* just in case its value is contained in $\{\mathbf{a}_g, \mathbf{d}_g\}$. Hence, according to $\mathcal{K}^5$ there are non-classical strongly assertible sentences, which explains why $\mathcal{K}^5$ can satisfy **AD** while it violates **MGBD**.

The paper is organized as follows. In Section 2, we state some general preliminaries. In Section 3, we give the formal definition of **MGBD** and **AD** and show how $\mathcal{K}^5$ testifies that these desiderata are not equivalent. Section 4 argues that **AD** is preferable over **MGBD** as a desideratum for theories of truth.

2 Preliminaries

L_T will denote a first order language without function symbols, with *identity* ($\approx$), a *truth predicate* (T) and with a *quotational name* ($[\sigma]$) for each sentence σ of L_T. L will denote the language that is exactly like L_T, except for the fact that it does not contain the truth predicate T. A *ground model* $M = (D, I)$ is an interpretation of L such that $Sen(L_T) \subseteq D$ and such that $I([\sigma]) = \sigma$ for all $\sigma \in Sen(L_T)$. A sentence may be denoted in various ways; $\overline{\sigma}$ will be used to denote any closed term, quotational name or not, which denotes σ in the ground model under consideration. We will make the simplifying assumption that a ground model has, for each of the elements of its domain, a constant symbol which refers to that element. This assumption has the advantage that quantification can be treated substitutionally so that we do not need to be bothered with variable assignments. With respect to $Sen(L_T) \subseteq D$ this assumption is unnecessary, as every sentence contains, per definition, at least one name: its quotational name. However, a sentence may also have a non-quotational name in a ground model, and this feature ensures that a ground model may contain self-referential sentences. Here are some notational conventions that we will respect in this paper concerning the use of some non-quotational names.

Definition 2.1. Some notational conventions. In this paper, the constants λ, τ, η and θ, will be used as follows, where I is an interpretation function.

1. $I(\lambda) = \neg T(\lambda)$. We say that $\neg T(\lambda)$ is a *Liar*.

2. $I(\tau) = T(\tau)$. We say that $T(\tau)$ is a *Truthteller*.

3. $I(\eta) = T(\eta) \vee \neg T(\eta)$. We say that $T(\eta) \vee \neg T(\eta)$ is a *Tautologyteller*.

4. $I(\theta) = T(\theta) \wedge \neg T(\theta)$. We say that $T(\theta) \wedge \neg T(\theta)$ is a *Contradictionteller*.

To be sure, the notational convention does not imply that every ground model contains one of the sentences just defined. However, if we use a sentence which is built with λ, τ, η, or θ, we always presuppose a ground model in which a Liar, Truthteller, Tautologyteller or Contradictionteller occurs.

As L_T is assumed not to contain function symbols, all the closed terms of L_T are given by its set of constant symbols, which will be denoted by $Con(L_T)$. Observe that $[\forall x T(x)] \approx [\forall x T(x)]$ is guaranteed to be a sentence of L_T. Given a ground model M, $C_M : Sen(L) \rightarrow \{\mathbf{a}, \mathbf{d}\}$ denotes the *classical valuation* of L based on M and is defined as usual[3]. Note that $C_M([\forall x T(x)] \approx [\forall x T(x)]) = \mathbf{a}$ and $C_M([\forall x T(x)] \approx [\exists x T(x)]) = \mathbf{d}$ for any ground model M. A *theory of truth* $\mathbf{T}$ takes a ground model as input and outputs a semantic valuation of the sentences of L_T. That is, $\mathbf{T}$ outputs a function $\mathbf{T}_M : Sen(L_T) \rightarrow \mathbf{V}$, where $\mathbf{V}$ contains *the semantic values of* $\mathbf{T}$. With $\mathbf{T}$ a theory of truth, $\top_{\mathbf{T}} = \mathbf{T}_M([\forall x T(x)] \approx [\forall x T(x)])$ and $\bot_{\mathbf{T}} = \mathbf{T}_M([\forall x T(x)] \approx [\exists x T(x)])$ are called the *classical top value* and *classical bottom value* of $\mathbf{T}$ respectively. Not any semantic valuation of the sentences of L_T qualifies as the valuation of a theory of truth. In this paper, I assume that in order for $\mathbf{T}$ to qualify as a theory of truth, $\mathbf{T}_M$ should *respect the world* and the *identity of truth*.

Definition 2.2. Theory of truth. Let $\mathbf{T}$ be a valuation method which, given a ground model M, outputs a valuation function $\mathbf{T}_M : Sen(L_T) \rightarrow \mathbf{V}$. We say that $\mathbf{T}$ is a theory of truth just in case, for every ground model M, we have that:

$$\forall \sigma \in Sen(L) : C_M(\sigma) = \mathbf{a} \Leftrightarrow \mathbf{T}_M(\sigma) = \top_{\mathbf{T}}, \quad C_M(\sigma) = \mathbf{d} \Leftrightarrow \mathbf{T}_M(\sigma) = \bot_{\mathbf{T}} \quad (13.2)$$

$$\forall \sigma \in Sen(L_T) : \mathbf{T}_M(T(\overline{\sigma})) = \mathbf{T}_M(\sigma). \quad (13.3)$$

That is, $\mathbf{T}_M$ should respect the world (13.2) and the identity of truth[4] (13.3).

Two interesting three-valued theories of truth are Kripke's *Strong Kleene minimal fixed point theory* $\mathcal{K}$, and his *Strong Kleene maximal intrinsic fixed point theory* $\mathcal{K}^+$. In order to define those theories, we let, for every ground model M, $\mathbf{FP}_M$ denote the set of all three-valued Strong Kleene fixed point valuations[5] over M. With $V_M, V'_M \in \mathbf{FP}_M$, we let:

$$V_M \leq V'_M \Leftrightarrow \forall \sigma \in Sen(L_T) : V_M(\sigma) = \mathbf{a} \Rightarrow V'_M(\sigma) = \mathbf{a}.$$

When $V_M \leq V'_M$ we say that V'_M *respects* V_M. The relation $\leq$ is a partial order on $\mathbf{FP}_M$. The following definitions are all taken from Fitting (1986). We say that V_M is *maximal*

---

[3]Modulo our use of assertible and deniable instead of true and false, which better fits in with the rest of the paper.

[4]Note that the identity of truth differs from the *intersubstitutability of truth*, according to which $T(\overline{\sigma})$ and σ are interchangeable in every (non opaque) context. In particular, revision theories of truth respects the identity of truth but not its intersubstitutability.

[5]We assume familiarity with the notion of a (three-valued) Strong Kleene fixed point valuation over M. To be sure, such a valuation has a Strong Kleene semantics, and it respects the world and the identity of truth.

just in case for no V'_M we have that $V_M \leq V'_M$, *minimal* just in case for no V'_M we have that $V'_M \leq V_M$. We say that V_M and V'_M are *compatible* just in case there exists a fixed point[6] V^*_M such that $V_M \leq V^*_M$ and $V'_M \leq V^*_M$. A fixed point V_M is called *intrinsic* just in case it is compatible with every other fixed point. For any ground model M, we let $\mathbf{I}_M$ be the set of all three-valued intrinsic fixed points over M. As Kripke (1975) shows, $\mathbf{I}_M$ has a maximum element and $\mathbf{FP}_M$ has a minimal element with respect to the relation $\leq$. Using the notions just defined, the definition of $\mathcal{K}$ and $\mathcal{K}^+$ is as follows. Let us remark that we think of the semantic values of $\mathcal{K}$ and $\mathcal{K}^+$ as given by the sets $\{\mathbf{a}, \mathbf{u}, \mathbf{d}\}$ and $\{\mathbf{a}, \mathbf{e}, \mathbf{d}\}$ respectively. When a sentence is valuated as $\mathbf{u}$, we say that that sentence is *ungrounded*, whereas a sentence that is valuated as $\mathbf{e}$ is called *extrinsic*.

Definition 2.3. $\mathcal{K}$ and $\mathcal{K}^+$. Let M be an arbitrary ground model. According to the theory $\mathcal{K}^+$, the valuation of L_T in M is given by $\mathcal{K}^+_M : Sen(L_T) \rightarrow \{\mathbf{a}, \mathbf{e}, \mathbf{d}\}$, where $\mathcal{K}^+_M$ is (obtained as) the maximum of $\mathbf{I}_M$. According to the theory $\mathcal{K}$, the valuation of L_T in M is given by $\mathcal{K}_M : Sen(L_T) \rightarrow \{\mathbf{a}, \mathbf{u}, \mathbf{d}\}$, where $\mathcal{K}_M$ is (obtained as) the minimum of $\mathbf{FP}_M$.

It is well-known that $\mathcal{K}^+$ respects $\mathcal{K}$, meaning that for every ground model M, we have that $\mathcal{K}_M(\sigma) = \mathbf{a} \Rightarrow \mathcal{K}^+_M(\sigma) = \mathbf{a}$. Not the other way around though: a Tautologyteller is strongly assertible according to $\mathcal{K}^+$ but ungrounded according to $\mathcal{K}$. That is:

$$\mathcal{K}^+_M(T(\eta) \vee \neg T(\eta)) = \mathbf{a}, \qquad \mathcal{K}_M(T(\eta) \vee \neg T(\eta)) = \mathbf{u}.$$

Definition 2.4. $\mathcal{K}^5$. The theory $\mathcal{K}^5$ is defined in terms of $\mathcal{K}$ and $\mathcal{K}^+$:

$$\mathcal{K}^5_M(\sigma) = \begin{cases} \mathbf{a}_g, & \mathcal{K}_M(\sigma) = \mathbf{a}; \\ \mathbf{a}_i, & \mathcal{K}_M(\sigma) = \mathbf{u} \text{ and } \mathcal{K}^+_M(\sigma) = \mathbf{a}; \\ \mathbf{e}, & \mathcal{K}^+_M(\sigma) = \mathbf{e}; \\ \mathbf{d}_i, & \mathcal{K}_M(\sigma) = \mathbf{u} \text{ and } \mathcal{K}^+_M(\sigma) = \mathbf{d}; \\ \mathbf{d}_g, & \mathcal{K}_M(\sigma) = \mathbf{d}. \end{cases}$$

$\mathcal{K}^5$ is a *Generalized Strong Kleene theory of truth* (*GS K* theory), whose semantics was discussed in the introduction. For the formal definition of the notion of a *GS K* theory—and the proof that $\mathcal{K}^5$ is a *GS K* theory—the reader is referred to Wintein (2011).

3 The Non-Equivalence of MGBD and AD

3.1 DEFINING MGBD AND AD

In this section, we define **MGBD** and **AD** rigorously. That is, we define the following three notions, the first two of which are taken from Kremer (2010):

- **T** dictates that truth behaves like a classical concept in M.

- **T** dictates that there is no vicious reference in ground model M.

- **T** dictates that all T-sentences are strongly assertible in M.

---

[6]Which we use here as synonymous with 'element of $\mathbf{FP}_M$'.

Definition 3.1. Truth as a classical concept. Let **T** be a theory of truth and M be a ground structure. A sentence σ is *classical*$_T$ in M, i.e., classical according to **T** in M, just in case $\mathbf{T}_M(\sigma) \in \{\top_T, \bot_T\}$. A set of of sentences is classical$_T$ in M just in case all its elements are. **T** *dictates that truth behaves like a classical concept in M* just in case $Sen(L_T)$ is classical$_T$ in M.

Definition 3.2. Strong assertoric pair. Let **T** be a theory of truth, let **V** be the set of semantic values recognized by **T** and let $\{\mathbf{x}, \mathbf{y}\} \subseteq \mathbf{V}$ such that $\mathbf{x} \neq \mathbf{y}$. We say that $\{\mathbf{x}, \mathbf{y}\}$ is a *strong assertoric pair* of **T** just in case for every ground model M and every $\sigma \in Sen(L_T)$:

- $\mathbf{T}_M(\sigma) = \mathbf{x} \Rightarrow \mathbf{T}_M(\neg\sigma) = \mathbf{y}$,

- $\mathbf{T}_M(\sigma) = \mathbf{y} \Rightarrow \mathbf{T}_M(\neg\sigma) = \mathbf{x}$,

- $\mathbf{T}_M(\alpha), \mathbf{T}_M(\beta) \in \{\mathbf{x}, \mathbf{y}\} \Rightarrow \mathbf{T}_M(\alpha \wedge \beta), \mathbf{T}_M(\alpha \vee \beta) \in \{\mathbf{x}, \mathbf{y}\}$.

With $\{\mathbf{x}, \mathbf{y}\}$ a strong assertoric pair of **T** and with $\mathbf{T}_M(\sigma) \in \{\mathbf{x}, \mathbf{y}\}$, we say that $\mathbf{T}_M(\sigma \vee \neg\sigma)$ is the *top value* of $\{\mathbf{x}, \mathbf{y}\}$.

Definition 3.3. Strong assertibility of T-sentences. Let **T** be a theory of truth, let M be a ground structure and let **V** be the set of semantic values recognized by **T**. Define $TOP(\mathbf{T})$ as the set of all top values associated with the strong assertoric pairs of **T**.[7] **T** dictates that all T-sentences are strongly assertible in M just in case:

$$\mathbf{T}_M(T(\overline{\sigma}) \leftrightarrow \sigma) \in TOP(\mathbf{T}),$$

whenever $\overline{\sigma}$ denotes σ in M.

What needs to be done in order to complete the definition of **MGBD** and **AD**, is to state Kremer's definition of a theory of truth dictating that there is no vicious reference in M. Before we present this definition, we sketch its rationale. If a set of sentences Y is classical$_T$, then the elements of Y *certainly* do not involve vicious reference. Let $Y \subseteq Sen(L_T)$ be classical$_T$. Then, the *sentential complement* of Y, $\overline{Y} = (Sen(L_T) - Y)$, consists of *potentially* problematic sentences, i.e., of sentences which may (or may not) involve vicious reference. Now if a ground model M can, intuitively, not *discriminate* between the members of $\overline{Y}$, i.e., if M cannot in any way discriminate between potentially problematic sentences, then, according to Kremer's definition, there cannot be vicious reference. M cannot discriminate between members of $X \subseteq Sen(L_T)$ just in case M is *X-neutral*.

Definition 3.4. X-neutral ground model and clean ground models. Let $X \subseteq Sen(L_T)$. A ground model $M = (D, I)$ is said to be *X-neutral* just in case:

- For each closed term $t \notin \{[\sigma] \mid \sigma \in Sen(L_T)\}$: $I(t) \notin X$.

- Non-logical predicates do not distinguish between elements of X. That is, with R an n place relation symbol ($R \neq T$) and with $d_1, \ldots, d_n, d_i' \in D$, it holds that if $d_i, d_i' \in X$, then: $(d_1, \ldots, d_i, \ldots d_n) \in I(R) \Leftrightarrow (d_1, \ldots, d_i', \ldots d_n) \in I(R)$.

---

[7]Thus, we have that $TOP(\mathcal{K}) = TOP(\mathcal{K}^+) = \mathbf{a}, TOP(\mathcal{K}^5) = \{\mathbf{a}_g, \mathbf{a}_i\}$.

A $Sen(L_T)$-neutral ground model is called a *clean ground model*.

Thus, in a clean ground model M, we can refer to sentences only via their quotational names and we cannot discriminate between sentences using any predicate in L. Intuitively, a clean ground model is a ground model with the 'least possible amount of vicious reference'. Here is Kremer's theory-relative definition of no vicious reference.

Definition 3.5. No vicious reference. Let T be a theory of truth and let M be a ground model. T *dictates that there is no vicious reference in* M just in case M is $\bar{Y}$-neutral for some $Y \subseteq Sen(L_T)$ which is classical$_T$ in M.

Observe that $\emptyset$ is trivially classical$_T$ in M for every theory T. Hence, from definition 3.5 it follows that **MGBD** and **AD** have the following corollaries:

MGBD corollary. If M is a clean ground model, then T dictates that truth behaves like a classical concept in M.
AD corollary. If M is a clean ground model, then T dictates that all T-sentences are strongly assertible in M.

In contrast to **MGBD** and **AD**, their corollaries are partially defined (only for clean ground models) and *theory neutral*, as the notion of a clean ground model is a theory-neutral notion. In a clean ground model, we cannot create Liar sentences, Truthtellers or any other sentences which are, intuitively, problematic. In a clean ground model there cannot, by definition, be any kind of vicious reference.

3.2 KREMER'S RESULTS AND THEIR CONSEQUENCES FOR $\mathcal{K}^5$

It can be verified, for arbitrary T, whether or not T satisfies **MGBD**. As we mentioned, Kremer did so for thirteen theories of truth—ten fixed point theories, three revision theories—in total. The next theorem summarizes Kremer's results.

Theorem 3.6. The results of Kremer
Consider five monotonic valuation schema's: Strong Kleene, Weak Kleene, Supervaluation and two schema's which are called $\sigma 1$ and $\sigma 2$ by Kremer (the details of $\sigma 1$ and $\sigma 2$ do not matter for our purposes). For each of these five schema's, define the associated minimal fixed point and maximal intrinsic fixed point, delivering a total of 10 fixed point theories of truth. With respect to the ten fixed point theories, we have that:

- The maximal intrinsic fixed point theory of any of the five schema's satisfies **MGBD**.

- Only the minimal fixed point theory of $\sigma 1$ satisfies **MGBD**.

Further, Kremer considers 3 revision theories of truth:

- T^*, the revision theory of truth based on stability, and T^c, the revision theory of truth based on stability and maximal consistency, satisfy **MGBD**. $T^\#$, the revision theory of truth based on near stability, does not satisfy **MGBD**.

Proof. See Kremer (2010). $\qquad\qquad\qquad\qquad\qquad\qquad\qquad\qquad\qquad\qquad$ □

We will be concerned[8] mainly with the results pertaining to the Strong Kleene theories $\mathcal{K}$ and $\mathcal{K}^+$. To show that $\mathcal{K}$ violates **MGBD**, it suffices to consider a clean ground model M_0. With LEM $:= \forall x(T(x) \vee \neg T(x))$, we have that $\mathcal{K}_{M_0}(\text{LEM}) = \mathbf{u}$, and so the result readily follows: there is no vicious reference in M_0 according to $\mathcal{K}$ and yet truth does not behave as a classical concept in M_0 according to $\mathcal{K}$. Similarly, the fact that $\mathcal{K}$ violates **AD** follows from the observation that $\mathcal{K}_{M_0}(T([\text{LEM}]) \leftrightarrow \text{LEM}) = \mathbf{u}$. Further, we have that:

$$\mathcal{K}^5_{M_0}(\text{LEM}) = \mathcal{K}^5_{M_0}(T([\text{LEM}]) \leftrightarrow \text{LEM})) = \mathbf{a}_i.$$

Thus, $\mathcal{K}^5$ does not dictate that truth behaves like a classical concept in M_0, the reason being that there are sentences, such as LEM, which are not classical$_{\mathcal{K}^5}$ in M_0, as they are not valued as $\mathbf{a}_g$ or $\mathbf{d}_g$. As there is no vicious reference in M_0 according to $\mathcal{K}^5$, it follows that $\mathcal{K}^5$ violates **MGBD**. On the other hand, from the definition of $\mathcal{K}^5$ and the fact that $\mathcal{K}^+$ satisfies **MGBD**, it immediately follows that:

$\mathcal{K}^5$ dictates that there is no vicious reference in $M \Rightarrow$
$\forall \sigma \in Sen(L_T) : \mathcal{K}^5_M(\sigma) \in \{\mathbf{a}_g, \mathbf{a}_i, \mathbf{d}_g, \mathbf{d}_i\}.$

Moreover, from the compositionality of $\mathcal{K}^5$, it follows that:

$\mathcal{K}^5$ dictates that there is no vicious reference in $M \Rightarrow$
$\mathcal{K}^5_M$ is *linear* (generalized) Strong Kleene w.r.t. $\mathbf{d}_g \leq \mathbf{d}_i \leq \mathbf{a}_i \leq \mathbf{a}_g.$

From the behavior of $\mathcal{K}^5$ with respect to $\mathbf{d}_g \leq \mathbf{d}_i \leq \mathbf{a}_i \leq \mathbf{a}_g$, it follows that a T-sentence of σ is valued as $\mathbf{a}_g$ when σ is valued as $\mathbf{a}_g$ or $\mathbf{d}_g$, and as $\mathbf{a}_i$ when σ is valued as $\mathbf{a}_i$ or $\mathbf{d}_i$. Accordingly, we get that:

$\mathcal{K}^5$ dictates that there is no vicious reference in $M \Rightarrow$
$\forall \sigma \in Sen(L_T) : \mathcal{K}^5_M(T(\overline{\sigma}) \leftrightarrow \sigma) \in \{\mathbf{a}_g, \mathbf{a}_i\}.$

Thus, if $\mathcal{K}^5$ dictates that there is no vicious reference in M, all the T-sentences will be strongly assertible in M. To sum up, we get:

Theorem 3.7. $\mathcal{K}^5$ **violates MGBD and satisfies AD**
Proof. Given above. □

4 Concluding Remarks

As mentioned in the introduction, Kremer cites Gupta (1982) with respect to the rationale of **MGBD**. The following quote of Gupta directly precedes the quote that was given in the introduction.

---

[8] As argued by Kremer, his obtained results put doubt on Gupta and Belnap's claim that revision theories of truth have, as a distinctive general advantage over fixed point theories, their "...consequence that truth behaves like an ordinary classical concept under certain conditions—conditions that roughly can be characterized as those in which there is no vicious reference in the language." (Gupta and Belnap 1993, p. 201). As the claim of Gupta and Belnap is cast in terms of the intuitive theory-neutral notion of no vicious reference, Kremer's results cannot be said to falsify their claim.

We conclude, then, that in a variety of circumstances we can consistently maintain the fundamental intuition. I suggest that it is a reasonable adequacy condition on any theory that purports to explain the meaning of 'true' that under such circumstances it preserve the fundamental intuition - or at least the intuition should be preserved if it does not come into conflict with some other intuitions that are of equal or greater importance.

Gupta 1982, p. 19

By *the fundamental intuition*, Gupta means the (strong) assertibility of all T-sentences, and *not* the behavior of truth as a classical concept. I agree with Gupta that a fundamental intuition concerning the notion of truth is the assertibility of all T-sentences. Also, I agree that it is desirable that a theory of truth respects our fundamental intuitions concerning truth when there is, intuitively, no vicious reference in M. Moreover, I agree with Kremer that a precise articulation of the intuitive notion of no vicious reference has to be theory-relative and I think that Kremer's theory-relative notion of no vicious reference is a satisfactory such notion. Accordingly, I advocate **AD** as a formal and theory-relative desideratum for theories of truth. **MGBD** is a *stronger* desideratum than **AD**: *if* **T** dictates that truth behaves like a classical concept in M, *then* all the T-sentences are strongly assertible in M, but not the other way around. This means that an advocate of **MGBD** needs an argument why **AD** is too weak as a desideratum for theories of truth. The rationale of both **MGBD** and **AD** is that our fundamental intuitions concerning the notion of truth should be preserved in situations without vicious reference. In order to argue that **MGBD** is preferable over **AD**, one thus needs to claim, or argue, that 'the classicality of truth' belongs to those fundamental intuitions. I, for one, do not want to claim this and also, I do not see how one could argue for this claim. As **AD** captures the (cited) rationale of **MGBD** and does not depend on the controversial claim under consideration, I take **AD** to be an improvement over **MGBD** as a desideratum for theories of truth.

References

Fitting, M. (1986). "Notes on the Mathematical Aspects of Kripke's Theory of Truth". In: *Notre Dame Journal of Formal Logic* 27, pp. 75–88.

Gupta, A. (1982). "Truth and Paradox". In: *Journal of Philosophical Logic* 11, pp. 1–60.

Gupta, A. and N. Belnap (1993). *The Revision Theory of Truth*. Cambridge, MA: MIT Press.

Kremer, P. (2010). "How Truth Behaves When There's No Vicious Reference". In: *Journal of Philosophical Logic* 39, pp. 345–367.

Kripke, S. (1975). "Outline of a Theory of Truth". In: *Journal of Philosophy* 72, pp. 690–716.

Wintein, S. (2011). "Generalized Strong Kleene theories of Truth and the Method of Closure Games". Submitted.

www.ingramcontent.com/pod-product-compliance
Lightning Source LLC
Chambersburg PA
CBHW070446090426
42735CB00012B/2477